W9-BRD-837

THE THREE HUNDRED YEAR WAR

Books by William O. Douglas

Towards a Global Federalism
Russian Journey
Beyond the High Himalayas
Almanac of Liberty
Farewell to Texas
Points of Rebellion
International Dissent
Holocaust or Hemispheric Co-Op

THE THREE HUNDRED YEAR WAR

A CHRONICLE OF ECOLOGICAL DISASTER

WILLIAM O. DOUGLAS

Associate Justice of the Supreme Court

RANDOM HOUSE
NEW YORK

Copyright © 1972 by William O. Douglas

All rights reserved under International and Pan-American Copyright Conventions.
Published in the United States by Random House, Inc., New York, and simultaneously in Canada by Random House of Canada Limited, Toronto.

Library of Congress Cataloging in Publication Data

Douglas, William Orville, 1898–
The three hundred year war.

Includes bibliographical references.
1. Ecology. 2. Man—Influence on nature. I. Title.
QH541.D68 301.31 72-3640
ISBN 0-394-47224-1

Grateful acknowledgment is made to Little, Brown and Co. for permission to reprint "Song of the Open Road" from *Verses from 1929 On* by Ogden Nash. Copyright 1932 by Ogden Nash. This poem originally appeared in *The New Yorker.*

Manufactured in the United States of America by The Book Press, Brattleboro, Vermont

98765432

First Edition

301.31
D737t

194227

CONTENTS

I AM INDEBTED TO:

Nan Burgess and *Fay Aull* for their oversight of this manuscript.

Linda Giese for her painstaking typing.

Dick Houston for his alert retrieval of governmental reports and documents.

Dagmar Hamilton, without whose research and patient, creative suggestions this book would not have been completed.

THE THREE HUNDRED YEAR WAR

THE THREE HUNDRED YEAR WAR

If only the clamour of words could die down in the world for a day and the modern man could be altogether silent, as the Bushman was commanded to be when he felt the tapping coming on in himself, he might hear perhaps his rejected aboriginal self, his love of life and element of renewal, tapping on his own back door.

LAURENS VAN DER POST,
The Heart of the Hunter

It is not accidental that nature itself was ascribed the feminine gender. Women, as nature, were mysterious, dangerous, volatile, and therefore something to be conquered, indeed, to be raped.

BARRY WEISBERG,
Beyond Repair

THE THREE HUNDRED YEAR WAR started in a modest way, because only axes and gunpowder were available. The wilderness was leveled in part to obtain fields for planting and to build towns. It was also leveled to obtain the great riches that came from the conversion of the eastern hemlock and great white pine into dollars. A messianic zeal accompanied this leveling, for it was part of our heritage to consider the wilderness as dangerous, if not evil, as a place filled with great hazards, which must, therefore, be laid low. That zeal extended from Plymouth Bay clear across the nation. It operates today as a force that confronts conservationists and ecologists, making them often seem un-American because they are against "progress."

The Three Hundred Year War included among its early victims the original inhabitants of the continent. They were the "heathen" who did not walk godly paths and were not imbued with godly attitudes. Yet they were, I believe, more respectful of the earth and its wonders than the newcomers whose creed and dogma showed little respect for the estuaries, sweet-waters, woods, and rivers. The leveling of the "heathen" went on relentlessly across this pleasant land. And when gunpowder and firepower had broken their organized resistance, they were still pursued. The covetous trader and speculator, yes and the politician too, euchred the Indians out of their choice lands and possessions, hounding them almost to extinction. When the Sioux hunted buffalo they killed only what they needed. But the white men sometimes took only the hides and sometimes only the tongues. "You can see that the men who did this were crazy," said the Sioux.

The Three Hundred Year War took an awful toll of our wildlife. Predatory man trapped and hunted without restraint. At first these resources seemed unlimited. But soon organized business, interested in the "fast buck," not in preservation of the wonders of this the most beautiful continent of the earth, began to specialize in extermination.

The early fur trappers played a major role in the movement West—they would go into a virgin territory, and then when settlers followed, move further West in search of game. As the fur trade flourished, wildlife diminished. Only the change in fashion of tall top-hats from beaver to silk saved any remnants of living beaver.

It was common to kill game birds—grouse, quail, woodcock, ducks, prairie chickens—and sell them on the market. Birds were slaughtered for plumage—the snow egret, white heron, birds of paradise, goura pigeons, skylarks, and most songbirds. Protective laws were finally passed. Meanwhile the carrier pigeons, once so plentiful as to blot out the sun as they passed overhead, were shot and netted and shipped by gunnysacks to the hungry markets of the East. The species at last was completely exterminated.

The emblem on the Great Seal of the United States—the bald eagle—represents a part of the nation that is dying. There are only about 750 breeding pairs left, due largely to DDT. While the bald eagle declines, the wild turkey (Benjamin Franklin's choice over the bald eagle) is getting more and more common. The prairie chicken became almost extinct; but heroic efforts are being made, particularly in North Dakota and Texas, to restore and preserve its native habitat—thanks mostly to private groups like Nature Conservancy. Private groups, such as the Audubon Society and the Wild Life group, not government, have been mostly responsible for saving disappearing species.

The drive against crocodiles and alligators started so that the demands of fashion could be filled. We may be down to two hundred crocodiles (all in Florida); and the alligator hangs on only precariously. The killing of alligators has greatly increased; some 127,000 were killed in 1968, 1969, and 1970. Most of those slaughtered were in Florida, and some 460 poachers were apparently involved. According to indictments returned early in 1972 most of the hides went to an Atlanta middleman who sold the hides to Japan, where they were used in making shoes, purses, luggage, and jewelry items.

As a result of murderous slaughters, the buffalo went. And as we moved West dozens of other birds and animals were practically wiped out. The prairie chicken was one; the blackfooted ferret another; the fisher, the ivory-billed woodpecker, the condor, and many others acquired the same distinction.

It looked for a while as if whales might survive the slaughter. When kerosene replaced whale oil as fuel for lamps, the early whaling industry collapsed. But new commercial uses, combined with modern methods of tracking the whales down, give them no escape. Whales now supply cat food, cosmetics, and transmission fluids (all of which are available from other sources); and they are hunted by radar-equipped catcher boats and helicopters too efficient to evade.

The search for the dollar drives the mightiest animals the earth has known into oblivion. They are supposed to be under the protection of the International Whaling Commission. But it is almost wholly dominated by the commercial whaling interests. The whales have few friends in court.

The main culprits are Japan and the Soviet Union, which together kill about 84 per cent of the 40,000 or more whales a year. The killings since the 1920's exceed all whale killings over the prior four hundred-year period. Factory ships now sail the seas; scout or killer ships and sonar locate the whales; the animals are frightened by high-pitched sounds and chased to exhaustion. The killer boats are fast, shooting 150-pound harpoons. The whale is towed to the factory ship and in an hour is cut into parts.

These days one whale is killed every twelve minutes somewhere in the world. Even though whales are close to destruction, the voices of those who would save them are not heard.

The ongoing attack also concerns the slaughter of baby harp seals so that their white furs can go to market. Each spring some 250,000 harp seals are clubbed to death so that their pelts can be made into coats, handbags, and luggage. These are mostly baby harp seals whose white fur lasts only from the age of three days to about seventeen days before turning gray.

In 1971 the Canadian quota was 245,000, though the Canadian and Norwegian hunters got only 219,000. Though Canadian biologists warned that the herd was approaching extinction, the Canadian quota for 1972 was 150,000. And that indicates that the herd is disappearing. There apparently is *malice aforethought* in the project, for the harp seal competes with fishermen and carries an obnoxious marine parasite. Behind the scenes are American fur and leather industries whose views are backed by spokesmen of our Commerce and our State Departments.

Next in line are the sea otters, transplanted in recent years to their ancient waters off California, but now relentlessly shot and stabbed by fishermen, who say the sea otter competes with them for the prized red abalone.

We have entered an era of "managed" wildlife that has serious impacts on our ecological balance. Thus the Park Service eliminated rattlesnakes in its Saguaro National Monument in Arizona. Why? To attract visitors. And why was that important? So that superintendents who served there would be identified with popular places and would be likely to be promoted. What was accomplished? A great increase in the kangaroo rat population that ate the young shoots and threatened the existence of the saguaros.

These historic accounts are not included in the present chronicle. I do not deal here with the age of axes and gunpowder with which we first despoiled this continent—the instruments that made China bald and turned the Middle East into a desert. This chronicle concerns the age of technology, deemed by many to promise life eternal, but which has become the quick engine of complete destruction, not only of wildlife but man himself.

The wild West is only a myth. Heavy smog washes against the slopes of the Cascade Mountains. Roads are everywhere and filled with the roar of jeeps and the small motorized "tote goats." Many waters where salmon and steelhead used to spawn are now only garbage dumps. Man-made badlands are being created by the new strip mines in the Far West. We are poor caretakers of the earth.

These forces are controlled here by the powerful drive for profits. In Russia the pressures come from the consumers, who want more to eat and wear, and from the bureaucrats, who strive for promotions and medals rewarding production skills. Though there are differences between the two countries, in each the supervisory bureaucracy is cold, paralyzing, and impenetrable.

That is indeed predestined, once Materialism and Technology become the Twin Gods.

Men are a violent lot—irrespective of race or color. War, not peace, has been their emblem. Every generation has had its war; and with the passage of time the violent nature of war has increased as a result of so-called technological advances. Our efforts in Vietnam and Russia's in the Middle East underscore the point. We protest against violence. Yet the violence with which we seem to be the most concerned is violence against the police, violence of the police against people, violence of people against people, violence of people against property.

Violence against our environment is another form of destruction that implicates our very survival, and it may well be a manifestation of deep subconscious forces. Our real alma mater is the Earth, without whom we are lost. Yet man's most devastating drives are acts of aggression against her. The Wintu Indians of California said, "How can the spirit of the Earth like the white man? Everywhere the white man has touched it, it is sore."

Many of our sanctuaries are fragile places. They could be trampled to dust by people who came reverently, especially if they came by the tens of thousands. That is true of Bryce, Zion, Mesa Verde, the Grand Canyon, and Canyonlands National Park in Utah. It is also true of our high alpine meadows where lichens and the bilberry grow. It is also true of the Guadalupes in West Texas that are waterless on top but which nurture a relic Douglas fir forest from the Pleistocene age. Yet the Park Service wants to pour the public into those high areas, where the fire hazards are enormous; and it wants to pave the lower canyon, where priceless botanical and zoological specimens exist. The National Park Service—like other federal agencies—services people, people, people. But the other members of the ecological community go largely unprotected.

When we resolve to preserve these sanctuaries as parks, we put them in real danger. They have been preserved to date only because very few people frequented them. But the desire to maximize visitors puts them in jeopardy. "They belong to all of us, don't they? Then we need roads to get into them. And how can a person be expected to enjoy them without the comforts of life?" So, here come the motels, roads, shops, dance halls, and restaurants that mark the demise of the sanctuary. Peter Parnall has written on and illustrated the theme in his book *The Mountain.*[1]

What we touch we are very apt to destroy. As a people we have no ecological ethic. We talk much about Law and Order and we mean it when we say that burglaries, street crimes, holdups and the like must cease. But in a deeper sense we have a basic disrespect of law—unless the law restrains the other group, not our own.

The corporate world—every pressure group—is always looking for means of avoiding or even evading environmental regulations. One of their techniques is to control the agency entrusted with protection of the public interest. This influence can be venal and sometimes is. The identity of interests between the regulator and the polluter may be subtle, not corrupt. It may be found in a common ideology that laissez faire is better than government control.

We have loosed a violent tide of destruction against most of the wonders of the wilderness. Cruel overgrazing of our alpine basins has converted them to deserts. The use of dynamite to destroy the redwoods wreaked havoc on the greatest of our native wonders. The power saw, the bulldozer, and the atomic bomb have now taken its place.

Much of this aggression against the biosphere is indirect. Madison Avenue wants to sell detergents guaranteed not to give housewives red hands and that instantly remove all stains. Phosphates seemed to be the answer. Yet phosphates from detergents became one of the great poisoners of our waters. In 1969 there were two billion pounds of phosphates in the four billion pounds of detergents sold in this nation.

After his hearings in December 1969, Congressman Henry Reuss of Wisconsin concluded that "the Interior Department is a branch office of Procter & Gamble" and the detergent industry was no longer "Mr. Clean." This subtle ruination of our clear blue waters continues largely unabated.

Substitute detergents without phosphorus had captured by 1971 about 10 per cent of the market. One of them had possibly cancer-causing properties. Others contained caustic substances posing a threat to children who might eat them or rub the granules in their eyes. But some phosphate-based products also contain caustic substances. The reasonable solution would seem to be banning the one that might cause cancer, labeling the ones with caustic substances so that users would receive fair warning, and, in stages, reducing the permissible phosphate content in all detergents so that shortly they would be phosphate free. Seven states require the phosphate content of detergents to be reduced, New York limiting it to 8.7 per cent in 1972 and to only trace amounts by 1973.

But the powerful detergent lobby, feeling the impending "crunch," reacted, brought pressure to bear, and got the federal regime to come out for phosphate detergents. Congressman Reuss replied that the protectors seemed more concerned about the health of the phosphate detergent industry than about the health of children.

Some of the destructive practices of modern technology

have unexpected and fatal consequences. Mercury poisoning, which I will discuss, is one example. The consequences of allowing mercury into our watercourses were not foreseen. The use of nitrogen fertilizers to boost corn production is another example. The nitrates that entered the soils were drained to the waterways and the nitrates became nitrites. These nitrites, when accumulated, reach proportions that are dangerous to infants, causing awful deaths as a result of methomoglobinemia. Barry Commoner has told the story in his recent book, *The Closing Circle.*[2]

We once grazed cattle. Now we fatten them in feed lots. The feed lots produce more organic waste than the total sewage from all our municipalities.[3] Our farm animals produce about two billion tons of waste each year. Once this manure was used as fertilizer for farm lands. Now commercial fertilizers are cheaper. The use and disposal of these wastes is perhaps our leading rural problem. The manure is largely converted to soluble forms of ammonia and nitrate which leach into the soil and reach the wells or are washed by surface water during rains into rivers and lakes. Runaway growth of algae is the result.

Cattle in feed lots are given DES—the growth hormone that speeds their growth and is so carcinogenic that several countries have banned it entirely. The hormone is excreted in the waste of the cattle in such concentrated amounts that it not only enters the watercourse—but does so in unbelievably high and frightening proportions. Moreover, cattle in feed lots are mostly fed grain, and the raising of grain induces the use of inorganic fertilizers (usually nitrogen) which heavily pollute the waters, as already noted. The intensive use of nitrogen enabled food production to increase while the acreage for farming decreased. But in the twenty years following 1949, five times the amount of nitrogen was used to produce the same unit of food.[4] That is another way of saying two things: (1) soil depletion goes on apace; and (2) the excess nitrogen used pollutes the waters.

So one must agree with the fertilizer salesman that his new technology is a success but only, as Barry Commoner puts it, "because it is an ecological failure."[5]

The environmental problem confronting humans has subtle connections with man's choice of violence to solve international problems and domestic problems, as well. Axes and plows are for amateurs. With bulldozers, high explosives, deep drilling, air-compression machines and the like, we have been able to speed up the destructive cycle enormously. We can now do in a decade the damage it took the ancients centuries to accomplish.

Our violence can be horrendous, as when we use nuclear bombs to dredge for us, or create huge craters that are radioactive for a thousand years by underground testing of our new nuclear warheads. It becomes quiet and unobtrusive, yet nonetheless lethal, when we pour industrial wastes into our waters.

We experience 10,000 oil spills a year, dumping 500,000 barrels of oil into the oceans. The world total is between one million and ten million *tons* a year, spilled or dumped in the oceans. Our leasing of offshore tracts goes on with speed and vigor under oil company pressure. The drilling now authorized off Louisiana can ruin an entire complex of marshes, estuaries, and barrier islands, indispensable for wildlife refuges. The decks are now being cleared for offshore Atlantic Coast drilling, which may well make the Santa Barbara disaster look minor.

More ominous, even than oil, is the menace of accidents involving chemical tankers. According to our Coast Guard, some 240 dangerous products are shipped in bulk by water. The hushed-up dieldrin accident in 1970 off the coast of Spain that wiped out the area's oyster industry is the forerunner of untold disasters yet to happen to the ocean.

The famous French oceanographer Jacques Cousteau predicts the oceans will be dead in fifty years. Jacques Picard, famous Swiss marine scientist, says they will have no life beyond twenty-five years. He says the first to die will be the Baltic Sea, the Adriatic and the Mediterranean the next.

A "dead" ocean is one that no longer produces the 55 million tons of seafood annually but, due to pollution, produces a decreasing annual crop. A "dead" sea is also one that produces a decreasing amount of oxygen. The planktonic diatoms in the ocean produce 70 per cent or more of the atmosphere's oxygen by photosynthesis. The toxic garbage, including DDT, which we have been putting into the ocean is slowing down photosynthesis.[6] Jacques Cousteau recently said, "Public opinion can save our dying oceans; but I believe the oceans will have a narrow escape."

If the so-called advanced nations keep on pouring chemicals into the environment without first testing their impact, we may produce air that is unbreathable and soil and waters that are poisonous. That must become Russia's, Germany's, Japan's concern as well as our own.

None of these actions is malicious in the sense in which we think of first-degree murder. But they are careless, thoughtless, and negligent, fitting more the category of manslaughter.

What we do marks a moral or ethical decline. We pass all blame on to the technology that creates these destructive devices. But technology has no value system: its existence depends upon creating masses of consumers.

What we do also marks the slow but sure destruction of the biosphere on which we are utterly dependent for survival. The enormity of our destruction is told by William H. Amos in *The Infinite River.*[7]

Overpopulation is often blamed for the present environmental crisis. Population of course affects living space; it tests the adequacy in size and quality of our recreational areas; it emphasizes how pitifully small are the roadless wilderness areas we have preserved; it underlines the need for wise land use if the urban sprawl is not to imperil the decencies of life on this continent.

Some say that people created the environmental problems

that technologists must now face; and that we must "stop blaming technology" for the fix we are in. But technology is our architect of disaster. The consumer is the victim of faulty carburetors on automobiles, of faulty smokestacks on furnaces, of leaking atomic energy plants, and so on.

The major problems confronting us on the environmental front, indeed, implicate largely technology; and they date mostly from the time when we became the leading gadgeteers of the earth.

Communist propaganda keeps saying that the world's dirty environment is due to capitalism, that only Marxism provides the political framework for maintaining a clean environment. But the facts are quite different.

The environmental conditions in Russia are as appalling as they are here. The Russian newspaper *Izvestia* recently reported:

"The Oka is an interesting river. If you go to a restaurant situated on its bank, not far from a perfume factory, you will be served a royal dish: carp with a 'rose' aroma, perhaps, or a pike with 'magnolia' scent. You can have still tastier dishes: perch, cooked in benzene, bream in kerosene or turbot in first-class lubricating oil. More than 380 large industrial enterprises dump their wastes into the Klyazma and the Oka. The Chernorechensky Chemical Plant alone dumps up to 150 tons of poisonous substances, including arsenic, ammonia, benzene and all kinds of acids and oil products. One hundred fifty tons a year or a month? No; per day."

Some of our rivers are now fire hazards. The Cuyahoga River and Houston Ship Channel have actually caught on fire.

Voltaire (1694–1778), while visiting in Russia, wrote home to Paris: "Today a letter from Peking did arrive. It took only six months. Incredible, isn't it? Progress goes forward with tremendous steps." Now our rate of change is very great. The problems of air and water pollution, the poisoning of fish, pheasants, and other wildlife, and the general problem of waste and littering date from about 1945, when we made revolutionary technological changes in our productive enterprises. During the same period there was an increased demand for electric energy; yet 90 per cent of it was due not to population increase but to the increasing number of electric eggbeaters, can openers, knives, and the myriad of other uses to which electric power is put.

Paper consumption has risen 100 per cent since 1950. But the consumption of plastics rose 759 per cent during the same time. Since World War II we have witnessed an onslaught of production, use, and discard. Every one of us discards about one ton of refuse per year and only about 8 per cent of the total is household garbage.

Barry Commoner[8] has computed that our production of articles conducive to debris since 1946 has shown the following percentage increases:

	°Per Cent
soda bottles	53,000
synthetic fibers	5,980
mercury (for chlorine)	3,930
mercury (for paint)	3,120
air-conditioner compressor units	2,850
plastics and resins	1,960
nitrogen fertilizers	1,050
electric housewares	1,040
synthetic organic chemical commodities	950
chlorine gas	600
electric power	530
pesticides	390
wood pulp	313
truck freight	222
consumer electronics	217
motor fuel consumption	190
cement	150

Food, textiles and clothes, household utilities; steel, copper, and other basic metals have grown at the pace of the population increase.

And consumption of the following ecologically sound items has not kept pace with population growth, or has actually shrunk:

	Per Cent
railroad freight	+17
lumber	− 1
cotton fiber	− 7
returnable beer bottles	−36
wool	−42
soap	−76

The Report of the President's Materials Policy Commission made in 1952 opted for "the principle of Growth," saying that "it seems preferable to any opposite, which to us implies stagnation and decay." People are beginning to wonder if "growth" may not be the long-range curse rather than the panacea.

Critics say that a "no growth" society would be repudiated by the poor as a conspiracy to rob them of the material things others have already acquired. Yet it is clear that the ravaging and raping of the earth that has gone on must stop. New amenities toward the earth must be shown by corporations as well as individuals. New amenities toward the less fortunate among us must be cultivated, as vast restructuring of society is needed. But there is no reason why the Good Life may not be enjoyed when we are in a state of equilibrium instead of in an era of roaring growth.

This book does not deal with the deadly napalm bomb nor the weapon we dropped on Hiroshima in a moral lapse, nor with the methodical use of herbicides in Vietnam to wipe out entire mangrove forests. We deal instead with the soft selling of Madison Avenue and the mass media that have made culprits out of housewives and that have put men who are pillars of the church only one jump ahead of the grand jury. It concerns stockholders who refuse to join proxy fights to make sure that the new slate of directors is opposed to poisoning our watercourses with mercury. It is also aimed at stockholders who silently revel in the profits from the "clear-cutting" of timber, even though salmon runs are destroyed and the watershed is cursed with the blight of a wasteland.

This book is a plea to all men to enlarge "the boundaries of the community to include soils, waters, plants, and animals, or collectively, the land," as Aldo Leopold put it, in his book *A Sand County Almanac.*[9]

It is the land ethic stated by the Sioux Indians through the centuries:

"With all beings and all things we shall be as relatives."

The Sioux talked of "the sacred hoop of the nation." While they did not know the chemistry, biology, and hydraulics of the ecosystem, they seemed to sense the indivisible quality of all life. "Everything the power of the world does is done in a circle."

AIR

Anything that becomes airborne, caught by the weather, is eventually brought to earth where it enters the environmental cycles that operate in the water and the soil.

BARRY COMMONER,
The Closing Circle

In this country we pour about 143 million tons of pollutants into the air in a single year. In thus using the atmosphere as a sort of garbage dump we unwittingly make co-conspirators of the weather and the sun.

J. L. MYLER,
Chaos & Conservation

ONE NEED NOT BE AN EXPERT to know air pollution when he sees it. In some areas the moisture of the eye mixed with smog produces a mild sulphuric acid which sets up an extreme irritation. The concentration of certain oxides in the air may be so severe as to cause women's nylons to deteriorate. And smog actually eats away granite, erasing in a few decades hieroglyphics that survived several thousand years of Middle East desert and sun. Los Angeles smog is also killing pines in the San Bernardino Mountains far to the east of the city. By 1970 it had killed 40 per cent of them.

The air of the inner cities is the worst air of all. The poor are exposed to carbon monoxide levels that are 30 per cent higher than levels experienced by the affluent. Technology can remove most of the irritants. Its application, however, costs money, and both industry and municipalities build their resistance on that issue.

Yet Florida in 1970 spent more to redecorate the Governor's office and House and Senate chamber than it spent on its air and water pollution control agency.

Jacksonville, Florida, owns an electric authority that spews sulphur from the stacks of one of its power plants. The city itself is responsible for about 80 per cent of the sulphur dioxide pollution in Jacksonville. It issues citations to private offenders but none of course to itself. New York City has been a notorious offender of its own smoke ordinances in the operation of its own incinerators.

The people who profit most from smelters in mining towns are usually shareholders who do not live there. The owners of the steel company in Birmingham, Alabama, live up north and do not want to do anything about the air in Birmingham.

Local industry's control over local smoke authorities is notorious. It is not a venal influence but one based on the desire to keep industry content and prosperous. But as the conditions worsen and the smokestacks continue to darken the sun, communities are beginning to move into action.

Those in the vanguard are Arizona, Illinois, Maryland, Michigan, Nevada, New Jersey, and Oregon.

More ominous in most respects are the automobiles and the planes and the smog they create. A federal law with ample power to impose controls now exists. The Clean Air Act of 1970 requires that in 1975 all new light-duty vehicles and engines reduce by at least 90 per cent emissions of carbon monoxide and hydrocarbons that were allowable in 1970, and that by 1975 they reduce emissions of oxides of nitrogen by at least 90 per cent of those allowed in 1971. But the new Act provides a procedure for a delay of one additional year beyond 1975 in making the new emission standards applicable to new cars.

The overriding question is whether the internal combustion engine can be made clean in that period of time. There are other alternatives, but the experts say the research and development necessary for a displacement of the internal combustion engine simply cannot take place on the 1975 timetable.

The 1970 Act gives the federal government control over emission standards of new automobiles and leaves the states control over emission standards for used cars. By the federal definition, a "used" car is one that has been driven for five years or for 50,000 miles; but the state standard can be more stringent or more lax than the federal standard.

To enforce the federal standard the Act gives the Environmental Protection Agency (EPA) administrator power to test motor vehicles on the assembly line; and if he finds that the new car will not meet the new standards he may revoke the certificate under which the manufacturer operates.

Each manufacturer warrants to the ultimate purchaser that the new car conforms to EPA standards and will comply for at least five years or 50,000 miles. If defects are found within that warranty period, repairs must be made by the manufacturer at no cost to purchaser or to dealer. The rub is that the five year or 50,000-mile provision does not come into effect until EPA decides that an adequate system exists for testing and inspecting the pollution-control device. But Congress set no date by which time the EPA administrator must make that determination.

The 1970 Act governs all types of air pollution, not automobiles alone. In its broader reaches it is concerned essentially with the emissions from industrial and municipal smokestacks.

The 1970 Act provides for federal air quality standards in ambient areas—which may be larger than one state or even smaller. The federal *primary* ambient standard will be based on what is "requisite to protect the public health." There is also a federal *secondary* ambient standard based on what is "requisite to protect the public welfare." The concept "ambient" averages out the air pollution over a designated area, meaning perhaps that in a particular spot the air may be chokingly dirty.

The states have nine months to submit plans for compliance with the primary standard and the secondary standard. These state plans for meeting the primary standards must be operative no later than three years, except, on application by a Governor, they can be extended for two years more. These state plans for meeting secondary standards must be operative within "a reasonable time" but can be extended for eighteen months.

While the federal plan is designed for an "ambient" area, the state plan will include emission standards, smokestack to smokestack. States are to establish emission standards for *existing* plants, which are subject to federal oversight. But where a factory, say, is involved, a Governor may apply for an extension of still another year.

When a *new* stationary source is constructed, it will be subject to federal emission standards. They become effective after ninety days but may be waived for "up to two years." Moreover, the President may grant an exemption for another two years; and thereafter he may grant additional two-year extensions ad infinitum. Apart from new autos, fuel additives, and aircraft, a state retains full control of air pollution, provided its control is not "less stringent" than the federal standard.

In April 1971, EPA promulgated national ambient air-quality standards[1] for sulfur oxides, particulate matter, carbon monoxide, photochemical oxidants, hydrocarbons, and nitrogen dioxide. In August 1971, EPA promulgated regulations[2] concerning state compliance with the federal standards. While 1975 was the statutory goal for clean air, it now seems likely that the goal will not be realized. For industry pressure was so great that EPA was forced to drop its key provision that industry be required to use the maximum avilable technology for controlling air pollution. So the battle lines will move from the federal sector to each state sector. Theoretically at least, a state can make its air-pollution controls as stringent as it likes.

Early in 1972, two states—Arizona and Montana—had stiffer smokestack standards than those provided by EPA for ambient air quality. Copper companies, the main complainants, were up in arms, pushing either to lower state standards or to make the federal standard preemptive. Ben Wake, Montana's air-pollution chief, replied that the federal standard would "make the country uniformly dirty." The copper companies in Arizona say they may have to close down unless more lenient state standards are adopted. The press is generally aligned with the copper companies; the public opinion polls show that the people want clean air. But as this book goes to press the Arizona health board has proposed relaxed rules and set them down for hearing. If the copper companies win this round and get the relaxed regulations, the question remains whether EPA will be forced to impose a stricter one.

While the costs of cleaning up the smokestacks of plants are usually stated in staggering terms of billions, it is estimated so far as electric power plants are concerned that state-of-the-art control would, at the most, add 10 per cent to monthly electric bills. And so the battle lines are drawn.

The temptation of course will be for industry to move into states like West Virginia, where there are few air regulations and which spew pollution into adjoining states. Loud complaints were, indeed, coming in 1971 from Christmas tree farms in Maryland that West Virginia generating plants were ruining their business.

The 1970 Act is severe looking. But when one reads it and the regulations closely, he soon realizes that it deals with air pollution very tenderly, allowing loophole after loophole for at least temporary escape. It was written to satisfy the utmost desires of the Establishment.

Though California had the power to put emission controls on existing cars, that smog-ridden state waited in vain for effective federal controls to come into force, because politics and technology stood in the way. The major polluters—and the manufacturers of pollution-prone cars and airplanes—acquired more valuable time to avoid a strict accounting.

A pollution-free engine for SST's has been found, but not one for automobiles and trucks. The barriers again are political and economic, with the fear that the imposition of new costs on present car owners would have powerful economic repercussions. Political pressures were also so potent that they eliminated the electric car and the steam car from possible competition with the autos of the well-entrenched petroleum industry. So the pressures mounted to keep the intolerable status quo in the saddle. The story is related more fully in the Nader Report *The Vanishing Air.*[3]

But Honda in Japan, it seems, has a rotary engine that offers pollution-free combustion of butane, propane, or methane gas. It is said to be revolutionary and to offer no pollution without the use of ponderous and costly anti-pollution equipment.

As this book goes to press, the pressures are on for easing the standards for nitrogen oxide emissions. This is the most difficult emission to control and probably the most damaging to growing plants and to humans. The propaganda is that the new restrictions will increase the cost of a car by at least $775; and it is beginning to become obvious that stiff controls make the cost of operating a car much more costly and also make it much more difficult to drive. It is becoming more and more apparent that a newer type of engine must be developed —quickly. Detroit seems to forget what the priorities are: first, clean air; second, consumer costs; third, profits for the manufacturer and for the oil companies.

Congressman Reuss recently said that the federal air-pollution program had been a thirteen-year failure. At Four Corners (Fruitland, New Mexico, where Colorado, Arizona, Utah, and New Mexico meet), at San Juan, New Mexico—near the Arizona and Colorado lines; at Navajo in Arizona, near the Utah line; at Mohave in California, near the Arizona line; and in Utah at Kaiparowitz and Huntington Canyon, coal-burning power plants are either operating or under construction which will supply far-off Los Angeles and other large centers with electric power. Four Corners operates under the aegis of Western Energy Supply and Transmission (WEST) Associates, a cooperative venture of some twenty-odd utilities and government power authorities.

The present plants have the virtue of being outside any metropolitan area; but they have the disadvantage of using low-grade and relatively cheap coal with a high fly ash, sulphur dioxide, and nitrogen oxide content. The entrepreneurs are not waiting for the arrival of devices to control these emissions, for pollution controls cost money. The plants will be polluters on a grand scale and their fallout will affect six national parks, twenty-eight national monuments, the Grand Canyon, Lake Mead and Lake Powell, and historic Indian lands.

One of these plants already belches more particulate matter than the cities of New York and Los Angeles combined.

It also pollutes the air of Albuquerque, some 150 miles away. When ten or twenty more plants are built, as is promised, the sacred blue sky of our unique Southwest will become a haze.

Senator Joseph M. Montoya of New Mexico said in protest, "We in New Mexico pride ourselves on having very fine air. We don't want to become an economic colony for any other state."

The Mohave plant takes ten tons of coal a minute, or five million tons a year. It requires 2,000 to 4,000 gallons of water a minute; and the water is Navajo and Hopi water drawn from wells over 2,000 feet deep. The water and coal are mixed and pumped as slurry through an underground eighteen-inch pipe nearly 274 miles long. It is a monument to modern technology and engineering, but a devastating depletion of water in a dry and dessicated land.

The water resources will be sucked dry. The rolling mesquite and cactus country will become rubble as a result of the strip mining we will discuss later. That is where our great-grandchildren will see the total destruction of an environment.

What price in defilement will we pay for this new source of energy? Is it worth all the frivolous "needs" to which we put electricity?

Three federal agencies, all in Interior, are responsible for air defilement in this Southwest power project: the Bureau of Indian Affairs, which leases the Indian lands, the Bureau of Land Management, which still sells coal to the various plants, and the Bureau of Reclamation, which is up to its eyes in conflicting roles:

(1) it has competing hydroelectric plants;
(2) it supplies water to cool the thermal power plants of WEST;
(3) it has water contracts with other WEST members which stipulate the design of the construction of the stacks emitting the smoke;

(4) it is the largest consumer of power from the Navajo Plant, using one-fourth of the kilowatts to be generated; and
(5) it is committed to water quality improvement, fish and wildlife enhancement, and outdoor recreation.

The bureau must be embarrassed each time it chooses which hat to wear. But the bureau must survive lest it shrivel up. The way to survive is to manufacture projects—one after the other. That is indeed what Parkinson's Law means.

That the three federal agencies should be the architect of a project that turns the clear sparkling air of the Southwest into smog is appalling.

WATER

This dismal little stream, when it has descended to the foot of the malign gray slopes, makes a marsh that is named Styx.

DANTE,
The Divine Comedy

A river is more than an amenity, it is a treasure.

New Jersey v. *New York*,
283 U.S. 336, 342 (1931)

Those who still fish in Lake Erie catch old inner tubes, beer cans, and other wastes of our use-and-discard society. The stench of garbage and sewage permeates the breeze in summer, and the ice freezes gray in the winter. Spring is truly silent there.

CECIL JOHNSON,
The Natural World

IN THE 1930'S AND 40'S we were thinking of water pollution as essentially a state problem. That is why Eisenhower vetoed a federal water-pollution control act in 1960. The federal government was making efforts to control oil pollution and to protect rivers as respects navigability. But as respects sewage and industrial wastes the federal government did little more than make loans or give grants in aid.

Most states had air and/or water pollution boards but the polluters generally controlled them. Some state laws allocated the board seats to industry, agriculture, and municipalities. That sent the goats to watch the cabbage. But whether or not there was a formal assignment of seats, the fact was that the polluters sat on most of the state pollution boards. That remained true in 1971.

The problem at the federal level has been "sticky," due to the heavy hand of industry at the controls. The National Industrial Pollution Control Commission is made up of 165 businessmen to advise the President on industrial-pollution reforms. Their work is organized through subcouncils. On the problems of pollution the Budget Bureau has one advisory council paid by these private groups. It provides reports on which new policies and new laws are based. But its meetings are private; consumer, environmental, and conservation groups are not present. Senators Lee Metcalf and John Moss and Congressman Reuss have led the fight against the federal agencies themselves creating advisory groups which are lobbies for special interests. Senator Metcalf said that they "enhance the corporate image," "create an illusion of action," and "impede officials who are attempting to enforce law and order."

The federal government has itself been one of our worst water polluters. Its seven installations on San Francisco Bay have been the notorious destroyers of our estuaries. Its ships —about one thousand in number, have no sewage disposal system, save for the USS *Canopus.* They deposit in San

Francisco Bay alone raw sewage equivalent to that of a city of 4,500 people. (Under a new Act passed in 1970, all vessels will in time have sewage treatment units—new vessels in two years and existing vessels in five years.)

The federal government, long responsible for Alaska, has already produced polluted rivers there. No city in Alaska has an adequate sewage treatment facility. Most major military bases dump their sewage raw into the inlets. Kodiak Harbor is so polluted from a seafood cannery that harbor waters can no longer be used in live holding tanks.

Waste disposal was a feature of the military installations made after World War II. But their design was not suitable for arctic conditions, and the problem of waste disposal has mounted.

Pulp mills are ruining the estuaries with their wastes.

Oil spills are common.

The pristine waters of Alaska are going the way of the Missouri River and the Connecticut River, in spite of all our experience and all the warnings.

Mercury pollution is not traced to industry alone. NASA daily infects Ohio's Rocky River with mercury.

First, as to industrial pollution, while we are far from a solution of the sewage problem, we are much worse off when it comes to industrial wastes.

The technology for handling industrial wastes—produced by factories run by those who have a mighty hand in Washington, D.C., affairs—is largely unknown. Those powerful people use municipal sewage systems which the taxpayers maintain; and their wastes often kill the bacteria which are essential for efficient operation of the sewage system.

Senator Metcalf of Montana reports that there are more than fifteen hundred advisory committees serving the federal government and that some of them indeed regulate the regulators. He points out[1] that the Industrial Pollution Control Council is made up of sixty-three top executives of major polluting companies and acts "as a massive, government-established public relations arm of the world's worst polluters." It and predecessor advisory committees are responsible for the nation's having no inventory of the composition, volume, and locations of industrial wastes in our waters. They have been called the "fourth branch" of government. "Working from a vantage point well within whatever administration is in power, they frustrate agency and citizen attempts to obtain the information needed to enforce the laws."[2]

No one knows (A) the chemicals which are entering our waterways from our industrial plants; or (B) the assimilative capacity of a particular waterway to absorb a particular waste without ecological harm, e.g., the safe amount of lead, mercury, iodine for each cubic foot of water; or (C) the extent to which the river or lake will, as in the case of mercury, serve as a synthesizing agency to convert a chemical from a harmless substance into a lethal one.

The toxicity of trace metals is well established; a dozen or more are suspect, as we shall see. Like mercury, they may be methylated by microorganisms in waterways to produce highly toxic substances that accumulate in fish and other sea products. Many of these trace metals were in man's customary environment. But such historical records as there are indicate that they were formerly present only in minute amounts. After 1940, however, with a great technological surge in chemistry, these toxics began to reach dangerous levels. An intensive scientific search is on to see what can be done about the vast tonnage of metals in our waterways and to learn how to avoid future accumulations.

The problem is not peculiarly American. It follows the progress of technology around the world, and has its lowest incidence in underdeveloped Africa. The task has worldwide implications, as all technologically developed nations contribute to the poisoning process that now implicates even the oceans. Collaborative efforts by international groups are now the first order of business.

On February 15, 1972, a modest start was made when twelve European nations (not including East Germany, Poland, and the Soviet Union) signed a convention to end dumping of poisonous wastes from ships or planes into the northeast Atlantic Ocean. Any sanctions will be imposed by the individual nations, not by an international agency. This convention is an important start, though ships and planes account for only a small part of marine pollution. It is indeed estimated that 90 per cent comes from industrial and domestic discharges through rivers, estuaries, outfalls, and pipelines.

To obtain control over industrial wastes, a monitoring system would have to be introduced by each nation to take into instant consideration all the variables I have mentioned so that when and if a danger point is reached, a particular discharge could be stopped or diverted to a holding basin.

No agency in this country makes any inventory of industrial waste discharges, though there have been outcries for a decade. The old poisons are not catalogued and the new ones, averaging fifty-five a year, remain unknown.

According to House reports, the force within government that prevented an inventory from being made was an advisory council, already mentioned, operating within the Budget Bureau. At last, in October 1970, under hammering of Congressman Chet Holifield and Congressman Reuss of the House Committee on Government Operations, a start was made toward making an inventory.

Congressman Reuss has said that the "present mercury crisis" could have been "largely averted" if industry working through the Bureau of the Budget had not prevented an inventory of industrial wastes being made for at least seven years.

At the present we grope in the dark and can only guess weeks after the event the reason why, for example, an enormous fish kill took place somewhere in the country.

Yet in 1968 the petroleum industry defeated a Muskie proposal that offshore rigs be liable for clean-up costs up to five million dollars for discharging oil into the ocean. The nuclear-electric power lobby defeated a proposal to require a federal license for power plants that caused water pollution, whether thermal or otherwise.

Water pollution killed an estimated 41 million fish in forty-five states in 1969—chemical poisons, thermal pollution, and nutrients from food products were apparently the largest killers, though no one knows for sure.

The case against the use of mercury—which pollutes the air as well as the water—is impressively strong, leading to the conclusion of experts that it should be banned. Even if that should happen tomorrow, the residue in the air and waters, in the birds and in the fish, and in the microscopic zooplankton of the ocean is so great that its curse would be with us for decades.

Under Congressional pressure the Army Corps of Engineers in 1970 began to issue permits for discharges of industrial wastes. But those permits related only to industrial wastes discharged into navigable waters.

On April 7, 1971, Congress finally authorized a full-scale control of industrial discharges into navigable streams. A permit system that was inaugurated dated from July 1, 1971, brought into coordination not only old Acts of Congress but the new ones. The latter include not only the one concerning water quality standards but also the National Environmental Policy Act of 1969, which provides that *environmental* as well as economic and technical considerations govern the decision-making of the Corps and other federal agencies. This new permit system touches only industrial wastes, not discharges into or out of public sewage treatment plants. The pessimists are now saying that a permit will be only a permit to pollute. Others are filled with nostalgia for the good days when enforcement was through use of a tiny, innocuous state club instead of the more lethal federal club.

Next, as to our free-flowing rivers, the ecological damage done by the Corps of Engineers[3] has been in part the destruction of free-flowing rivers and the bottom lands that teem with wildlife and abound with botanical wonders.

The Bureau of Reclamation built dams that made possible the watering of deserts. Over the years it has contributed greatly to the cause of irrigation. But now it too is propelled by Parkinson's Law to build more and more dams. Come with me to the Far West and I can show you its "white elephants." The bureau over the years served the nation well. But the Nader Report *Damming the West* (1971) shows that only Parkinson's Law is a reason for keeping the Bureau of Reclamation alive: (1) over 50 million acres of cropland are idle as a result of federal food policies; (2) another 110 million acres of cropland have not been retired as a result of federal efforts, yet remain idle; and (3) about two-thirds of the three million existing farms will not be needed in the long run.

The TVA engages in the same destructive practices in Tennessee and North Carolina. A witness at the Muskie Hearings in 1970 testified:

"We have, in our own Tennessee Valley, the destruction of thousands of acres of valuable farmland inundated behind dams with that silly excuse that waterways are needed to provide barge transportation so that industry will move into the area. We accept without question engineering studies that show us we need to dig another hole, build another dam, or another power plant because we need more power to process more iron, to build more automobiles, stamp out more tin cans, produce more bottles, manufacture more plastics, produce more paper so that we can use them for a short period of time, throw them away, and start all over again."

The TVA has become remote, seemingly hostile, opinionated, and anti-environmental. Dealing with it, say the people, is almost like dealing with a foreign power.

TVA, like most federal agencies, does not like the requirement of the National Environmental Policy Act of 1969, which requires a statement of the environmental impact of its projects—the adverse impact as well as that deemed to be beneficial. The TVA resisted making such a statement for a project started before the 1969 Act became effective, but it was brought to heel by a courageous federal judge in Tennessee, Robert L. Taylor.

The TVA has a board of directors named by the President and confirmed by the Senate, each to serve for a term of nine years. TVA has broad powers to launch projects and acquire property, those powers including the power of eminent domain. It must, of course, go to Congress for the appropriations for any project; and it is only at the level of the Appropriations Committee that the citizens can be heard. In practical effect TVA is a law unto itself and typical of the regime in power under socialism. A study of the operations of TVA reveals that socialism is no cure-all. The evils spawned by TVA are different in kind from those that flourish under private enterprise. The remedy for each is public participation so that the problem may be aired at an early stage and the course of decision altered. Socialist TVA has such an ugly demeanor that farmers whose land is sought for another needless dam are beginning to wait with rifles for the arrival of the TVA agent.

Hydroelectric dams were once the fad; and they went up with little regard for environmental consequences. Seattle City Light built Ross Dam on the Skagit some forty years ago and it was highly acclaimed. The demand now exceeds the supply and Seattle wants to raise Ross Dam. But the state of Washington is up in arms. So is Canada where the Skagit River rises. Ross Dam, like TVA, is a form of socialism. But whether a dam is part of a socialistic or a free enterprise regime, the environmental impact is the same; and as a people we are now insisting on making environmental standards our important guidelines.

Dams are often advertised as methods of flood control. Real flood control, however, comes from undisturbed valleys. If valleys are farmed or logged, proper farming or logging methods are the answer, not dams. Dams silt in—some in ten, some in twenty, some in fifty years—and become useless. Dams—with rare exceptions—do not fit ecological standards.

The Bureau of Reclamation limits its estimated *benefits* and *costs* of any of its dams to irrigation, hunting, and fishing, not to the preservation of unique species or other ecological factors.

The Aswan Dam on the Nile is already a scourge. Without the Nile's sediment, Egypt's downstream lands already need fertilizer; furthermore, costs of fertilizers are reducing by a fifth the income of farmers making perhaps seventy-five dollars a year. The new lands irrigated are negligible but those irrigated are now filled with snails that carry the disease bilharzia, known medically as schistomiaris, a great leveler of those who must work in water.

The Army Corps of Engineers, with its much-touted flood-control plans, is contributing to the death of the second largest American swamp—Atchafalaya in Louisiana. The swamp was nourished by overflow waters, some of which no longer reach it. Oil spills from wells being drilled in the swamp have hurt its wildlife. Pesticides have injured its fish and wildlife resources. The ducks have mostly disappeared. Lumber companies have engaged in heavy cutting. As the waters recede farmers convert the old wetlands into soybean farms. But with the mounting population the old swamp is becoming an attractive "wilderness" to many people. How can it be saved from the Corps and from "civilization"? Here again it is the federal government that is the great destroyer.

The Corps has 40,000 civilian employees who must be kept at work. Many ask, Why do they not build sewage disposal plants? That would require new statutory authorization and it is being pressed. I have seen lists, enthusiastically prepared, which name the generals after whom these sewage disposal plants could be named. The Corps produces tables showing *costs* and *benefits* from the project. But it never lists as a *cost* the destruction of a river and all of its wildlife. Its Benjamin

Franklin Dam on the Columbia would have wiped out one-third of the fall Chinook salmon, for example. But that was not listed as a *cost* by the Corps. But if dams must be built to satisfy Parkinson's Law, why should they not be built pursuant to ecological standards?

Salmon and steelhead trout that return from the ocean to spawn in sweet-water streams were prospective casualties of every dam built on the West Coast. Dams on the Columbia, where possible, contained fish ladders enabling the anadromous species to go hundreds of miles upstream. But another risk was not contemplated: that nitrogen is trapped deep in the pools by the long fall of water over the spillways. A fish which breathes in this nitrogen-saturated water is stricken, for the gas goes into its bloodstream; and when it rises near the surface the gas bubbles out, causing blisters beneath the skin, blinding the eyes, or even pushing them from their sockets. The fish is either killed or so weakened that it shortly dies. In 1970 most of the steelhead and most of the Chinook salmon migrating to the ocean were killed in this way in the Columbia. We still may pay an awful *cost* for our elegant Columbia River dams.

The damage done by dams is untold. It may delay the spring runoff long enough to drown out hundreds of duck and goose nests. It may alter the spawning cycle of trout and interfere with the natural insect reproduction so essential to trout. The reservoir behind the dam may provide irrigation water; yet often in the West it is used on poor land that is "mined" to raise potatoes or grain, the potatoes then being burned and the grain stockpiled.

Some dams wreak havoc on fish. Industrial dams on the Willamette, a tributary of the Columbia that comes in at Portland, Oregon, have killed tens of thousands of salmon and steelhead trout. The state of Oregon estimates that two such dams killed nearly six million of steelhead and salmon under fifteen inches as they were migrating downstream to go to the ocean.

Salmon feed on needlefish. A pulp and paper mill at Kitimat, British Columbia, has caused dead needlefish to pile up two feet deep on the beaches. Salmon are disappearing from that area—and cod too.

Beyond that is the damage caused when a riverbed is bulldozed and turned into a flume and when herbicides are used on the spoil banks to keep a new river bottom from returning.

The loss of the river is permanent. Along with it go the raccoons, badgers, fox, deer, and other wildlife that frequent the river bottom. Moreover, channelization of part of the stream can spell disaster to down-river portions. For when high water comes, the new flume pours torrents of water into the lower river bed, washing and gouging it out.

Exhibit A of the downstream damage done may be seen by visiting Trinity River at Fort Worth, Texas. Exhibit B is the monstrous concrete ditch which the Corps built to take the place of Tamalpais Creek in Kentfield, Marin County, California. The purpose of preventing flooding was not served. Moreover, work of that character is at war with the preservation of our flood plains—exciting stretches of lowlands which should *never be zoned for permanent occupancy.*

Channelization has had disastrous effects particularly on the lower Mississippi. Islands which once had unique recreational potentials have disappeared. So have wetlands, essential for bird life. Substantial acreages of public waters have been lost. Thus wing dams built by the Corps cause accretions that add to private riparian ownership and replace public waters. Channelization has caused the Missouri River to lose 80 per cent of its wet areas and its public waters.

The Corps' most recent proposal would divert the Platte River in Nebraska for over sixty miles, wiping out the riparian habitat critical to sandhill cranes, whooping cranes, ducks, and geese. This permanent ruination of the river's ecosystem summoned Nebraska's conservationists to action in 1971.

The Soil Conservation Service—once the darling of conservationists—now turns rivers into flumes, destroying the varied life of the river bottom and dangerously imperiling downstream private property. SCS now competes vigorously for the position of Public Enemy No. 1. It was so successful in the thirties and forties in showing farmers the advantages of hedgerows and the virtues of contour plowing that it was practically out of business. So, searching for a means of bureaucratic survival, it mimicked the Corps and got some flood-control projects from Congress, and greatly increased its jurisdiction under Eisenhower. When restraints on federal spending were lifted, it went into big projects and fairly blossomed under Johnson. SCS has more than three thousand conservation districts behind it, and they are represented in Washington, D.C., by a powerful lobby. SCS is allied with local construction and development companies which turn a pretty penny and receive a chance to build on what was previously marshy land. The farmers generally gain by picking up acreage of cultivable land. The losers are the taxpayers, the wildlife, and the entire community of botanical and zoological wonders represented by a river course.

SCS has 2,000 miles of streams in Alabama alone headed toward destruction. In Georgia it has 176 channelization projects which will affect every major fishing stream. As of April 1971 it had completed 284 projects in various states and approved 1,033 additional projects. It aims to channelize 11,000 streams, substituting sterile ditches for pleasant, meandering streams rich in fauna and flora. Like TVA, it fights and resists complying with the National Environmental Quality Act of 1969, which requires statements of the environmental impact which projects of a federal agency will have. It has destroyed more Maryland wetlands than all other causes combined—11,960 wetland acres of the total 23,717 lost during the 1942 to 1968 period. SCS can easily destroy all of Maryland's wetland with a few more channelization projects.

The same struggle goes on in Russia. The Minister of Agriculture usually pleads for saving the bottom lands from dams, while the Minister of the Power Industry talks about the need for "progress" and "development." The Minister of Agriculture answers by saying that additional electric energy should be supplied by conventional fuels plus nuclear energy. The Minister of the Power Industry points out the vast recreational potential of the reservoirs which he will build, sounding very much like our Bureau of Reclamation, our TVA, and our Corps of Engineers. And so the argument goes—around the world.

Under federal law the states must establish water quality standards. When a state does not establish water quality standards that are acceptable, the federal government can do so.

James Ridgeway[4] has told how under Stewart Udall, when he was Secretary of Interior, the water quality regulations actually lowered permissible standards of water in the pure streams with the result that the federal government was licensing pollution. Under pressure, Udall adopted a "non-degradation cause," which theoretically was designed to pledge the states not to lower existing water quality. But there was a joker in the clause—existing standards could be lowered if "such change is justifiable as a result of necessary economic and social development." Thus every polluter found an escape hatch.

Under Acts of Congress the federal government has jurisdiction over pollution of a waterway that occurs in one state and appears downstream in another. If, however, the pollution occurs only in one state, the federal government, without a request by the Governor of the state, has no statutory authority, as the laws are written, to abate the nuisance even though the pollution affects navigable waters. This statutory restriction on federal enforcement, without request by the Governor, has had a serious impact on Puget Sound, San Francisco Bay, Lake Ontario, Lake Huron, the Mystic River and the Charles River in Massachusetts, the James River in Virginia, the waters of Puerto Rico, the Virgin Islands, Hawaii, and many waters in Alaska.

The main federal enforcement device is a "conference" where state and federal officials meet. The conference device has produced occasional advances; but as Senator Muskie recently said, the conference has become less and less an enforcement device and more and more a discussion group.

The federal water program is mostly public relations. Little progress has been made. The laws favor the polluters, with the result that our waters get fouler with each passing year. Monopoly power is a kind of private government, which may exercise greater power than any of the public officials.

"If the steel companies of Indiana and Ohio decide they will use rivers as open sewers, the rivers die . . . If petrochemical companies in New Jersey pour filth into Raritan Bay and choking fumes into the air over Elizabeth, Perth Amboy, and Staten Island, who is to stop them?"[5]

Minnesota, in 1970, led the way in zoning the shorelands of its public waters, but ohter states lag far behind.

Beautiful Lake Tahoe seems destined to become an awful slum, its clear blue waters filling with algae due to the runoff of man-made silt. The interstate planning agency, which largely reflects the gambling industry and the local commercial interests, puts tremendous pressures on city and county authorities to raise the tourist influx from 35,000 to 150,000, to line the shores of the once-sapphire lake with more casinos, to build towering hotels and mushroomlike condominiums, to construct shopping centers combined with residential units, to lay out vast home sites first and then to look for water and plan sewage disposal last.

Lake Tahoe—famed for its crystal blue waters and its magnificent stellar jays—seems destined for destruction. The people who go there do not even know the glories of a sunrise on a spit of white sand washed by sparkling water. They come to Nevada only to gamble and defile one of God's most startling sanctuaries.

Lake Tahoe can be saved only if the federal government, which after all owns the basin, directs all federal agencies there to adopt stern environmental measures and to send gambling and the sleazy development it generates out of the basin.

Vermont has instituted a permit system under which a factory must agree to meet certain water quality standards in order to discharge its wastes into a river or lake. Those who cannot meet the water quality standard can get only a temporary one-year permit for which they pay a fee estimated to be the cost of restoring the stream or lake to the necessary standard. The permit is temporary because of the expectation that the industry will find the technology necessary for full compliance.

In contrast to enlightened Vermont is Texas, where the cost of pollution is never borne by the polluter.

Texas authorizes the discharge of brine as well as oil into deep wells—the fresh water acquifiers that underlie most of the nation. One gallon of brine can contaminate 130 gallons of fresh water for drinking purposes. In Texas, about 29 billion gallons are released annually either in surface waters or pits. One barrel of crude oil escaping into water will contaminate 10,500,000 gallons of fresh water, making it unfit for fish life. How much brine and oil escape from wells into percolating waters is not known; but when contamination is evident, it is too late to do anything. Indeed, once the brine or oil escapes, it may effectively sterilize large acreages permanently.

In Texas all toxic metals—mercury, arsenic, barium, boron, cadmium, copper, chromium, lead, etc.—may generally be discharged in waters, provided a permit is obtained. The secret way in which permits for such discharges are issued by the Railroad Commission is related in the *1971 Joint Report of the Interior Committees on Pipeline Study and Beaches.* No matter the degree of harm done, there are no penalties unless there is no permit or unless the discharges are in excess of the volume authorized. What underground network of water connects with a given well no one knows. We may have already greatly imperiled our percolating waters for customary domestic use.

There are in other words no controls—either state or federal—over underground waters and their pollution, both by industrial wastes and by sewage.

The problem of ground waters is already acute in some areas. On Long Island they are the sole source of supply for at least one and a half million people and several thousand commercial enterprises and industries. In Georgia a few industries take most of the ground water—65 per cent. The massive suction in the Savannah area has resulted in salting wells fifty miles away. Many ground waters are nonrenewable, being trapped water from ancient glacial times. In Lubbock, Texas, ground waters not only supply the town folks but also the irrigation needs of farmers.

Even in remote, rural areas, well water is often contaminated from septic-tank sewage, with the result that entire families are the victims of amoebic afflictions for the better part of their lives. Local regulation is almost nonexistent. In Door County, Wisconsin, it was recently discovered that 110 of 457 private wells were contaminated and that 15 per cent of public accommodations served polluted water.

It is estimated that in Arizona ground-water levels have dropped 720 feet since 1900. Hence the Pacific Ocean is seeping into the underground of that area; and the Gulf is seeping into Texas. And even in the rainy area of Long Island, man's draining of ground waters has caused the Atlantic to seep in.

It is common for industry to drill holes about six inches in diameter to depths of 3,000 to 12,000 feet and through pressure force toxic fluid wastes into porous sandstone or limestone. It was disclosed by Senator Thomas F. Eagleton of Missouri in 1971 that industries, in order to cut down pollution of waterways in Louisiana, drilled sixty-two deep disposal wells. No one knows what percolating waters, what artesian wells are now poisoned beyond redemption.

Another hiatus in our water program is the lack of control over discharges of sewage into coastal waters. The ocean is everyone's cesspool—the world around.

Finally, as to sewage, long delays in building adequate sewage treatment plants, a burgeoning population, and the critical fiscal problems of municipalities and states have resulted in most of our rivers being open sewers and many of our lakes becoming cesspools. While the appropriations have mounted, we were, overall, further behind in 1972 than we were in 1957.

Secretary of the Interior Walter Hickel, who to the surprise of many was an aggressive conservationist, estimated in 1970 that it it would cost $10 billion to bring all municipal treatment plants into compliance with the new water quality standards so that all waters—except the Great Salt Lake—would once again be fit for human use, swimmable though perhaps not drinkable. We spent $10 billion in less than five months in Vietnam. What order of priorities do we really have?

The federal government is said to "keep the word of promise to our ear and break it to our hope." The promise of the federal government to act has indeed often proved illusory. The federal government made big and bold proposals for water quality standards. For 1968, $450 million was authorized but only $203 million appropriated; for 1970, $1 billion authorized, $214 million appropriated.

The Federal Water Quality Administration set 1972 as the goal for creation of a clean Lake Erie, the lake known around the world for its pollution. But in 1971 that goal appeared to be wholly illusory: 78 of 110 cities were not meeting a clean-up schedule set in 1967; 44 out of 130 industries were behind schedule; and the federal government had provided only 10 per cent of the funds promised.

Though the technology of sewage treatment is fairly well advanced, there are unsolved problems. The effluent has a high phosphate content and phosphates in a river or a lake may cause eutrophication, changing blue waters to algae-infested waters. Devices are known for removing the phosphate in usable form. But the processes have not proceeded beyond the pilot plant stage.

Sewage plants may render water disease-free, but they do not remove the excess of nutrients which is present in sewage and is so damaging to fish and other aquatic life. Though this sounds paradoxical, these nutrients stimulate so large a growth of plant life that the oxygen in streams is depleted. The truth is, the nutrients could be removed at costs less than those involved in our budget for exploring the moon. But though water is critical in the maintenance of the ecological system on which life is dependent, we give the moon priority.

The sludge from sewage is today commonly burned. There are known ways to recycle it and make it returnable to the soil as a fertilizer. But that process is practically in its infancy.

Under existing attitudes, regulations, and forms of financing we are far, far away from solution of the sewage problem.

Some say we turned the corner in 1971. Despite its previous poor record in ecology, the Corps, which has rightly been said to have been "instrumental in almost every great forward step in American destiny," in the spring of 1971 entered into an agreement with the Environmental Protection Agency—the federal agency of oversight—to undertake field studies of alternative methods of modern sewage disposal in five of the nine largest metropolitan areas in the nation. Some millions of dollars have been shifted to the project from previously budgeted Corps funds.

The technology exists to give our waters maximum, not minimum, protection. The brains have at long last been drafted to start the job.

The Corps has already made public the Chicago phase of its study. It offers three ways of treating metropolitan Chicago's 1.2 billion gallons a day of combined industrial and sewage wastes:

(1) the advanced biological method now in vogue which upgrades existing sewage treatment plants;
(2) a new "physical-chemical" system of plants along our waterways;
(3) the "land-treatment" method which includes "spray irrigation" systems.

The Chicago Report gives an economic analysis of the three systems:

Capital Costs. "Advanced biological"—$3.64 billion (not counting investments in existing plants); "physical-chemical," $2.99 billion; "land-treatment," $2 billion.

Capital costs per Gallon of Installed Capacity. "Advanced biological," $1.53; "physical-chemical," $1.26; "land-treatment," $.84.

Local One-Year Operating, Maintenance, and Replacement Costs (per capita). "Advanced biological," $.26; "physical-chemical," $.25; "land-treatment," $.09.

Spraying sewage on the land is not new. It has been used in California with some success; and in 1972 Vermont authorized ski areas to dispose of their treated sewage by spraying it on the ground.

The "land-treatment" proposal of the Corps will probably carry the day, as it will completely halt the industrial and municipal pollution of our waters and do so at a greatly reduced cost. Thrift and sanity are still good American hallmarks.

The ultimate solution of the water-sewage problem will also include the flush toilet, an article to which we are now as addicted as we are to coffee. The flush toilet uses 99.9 per cent of pure water to get rid of a tiny per cent of pollution. Each person uses a hundred gallons of water a day, as compared with ten gallons one would need if privies were used.

The enormous cost in water is one price paid for the flush toilet. The other is the loss to the soil of organic matter.

The privy would answer the water problem; but it is clearly impractical in urban centers. Such a revolutionary program would entail institutional changes of profound significance and would presuppose a public mood of acceptance of a project that has explosive connotations.

The oncoming critical problem of the nation and of the world is potable water. Devices for saving—or recycling—water must be designed.

Nevertheless, technology is on the move. Britain has flush toilets that save 70 per cent of the water used by standard American models. Sweden has one that uses air rather than water. And experiments in this country for on-site disposal of solid wastes are underway in sparsely settled areas. The elimination of—or vast reduction in the use of—flush toilets is the heart of the sewage problem.

The latest chapter can be graphically summarized by a report on the Mississippi River and on the Muskie bill as it passed the Senate.

By 1971 the Mississippi—the father of waters and our greatest river—was in danger, and so was aquatic life in the Gulf of Mexico, as well as human life in the lower reaches of the river. Industry, municipalities, and agriculture pour tons of poisons into this mighty stream. EPA found forty-six organic chemicals in the drinking water of New Orleans. There were eighty-nine organic chemicals in the waters as they reached the Gulf. Lead, mercury, cyanide, phenols, arsenic, copper, chromium, zinc, and cadmium impregnated the river. Nearly four hundred communities with a total population of 2,370,000 poured raw sewage into the Mississippi or its tributaries.

Late in 1971 the Senate passed a bill designed to "restore

and maintain the natural chemical, physical, and biological integrity of the nation's waters." It aimed to bar the discharge of pollutants in navigable waters by 1985, to attain by 1981 "an interim goal of water quality" which fish, shellfish, and wildlife as well as man would find tolerable, and to ban the discharge of toxic pollutants into all waters. It provided for the use of "the best practicable waste treatment technology" before the discharge of waters; but in case existing technology does not eliminate the pollutants, the proposed Act will provide for the application of any new technology at a later date.

The Muskie bill, though strong, has many weaknesses. The permit regulations issued by states have no "conflict of interest" provisions, making it easy for polluting companies to man the state agencies. There is no mandatory federal review by EPA of state action. Revocation of permits once granted has no teeth, for it depends largely on negotiation. Finally the Muskie bill, while granting an individual the right to complain, places on the individual all of the financial risks of litigation, while existing law places that burden on the state and rewards with a fee the complaining citizen whose complaint wins out.

A stream of criticism from the Administration rose against the bill in the House. The timetable in the Senate bill, it was said, would be much too costly and the combined demand from industry and municipalities for waste treatment plants would outstrip the capacity of the construction industry to build them.

The status quo, controlled by the polluters, is struggling to stay in power and to perpetuate a system that produces polluted waters for those yet unborn.

The Chicago Report of the Corps will doubtless have a great impact on the Hill, where the battle for and against the Senate bill was shaping up as this book went to press. The "land-treatment" process recommended by the Corps will in time relieve our waterways of the curse and burden of industrial and sewage pollution—at less than two dollars a month per person!!

The catch phrase *full development* has long been the presupposition of river basin planning. It expresses our preoccupation with technology and engineering. That approach must give way to ecological management.

The crisis is not merely environmental. It is the crisis of the survival of life on earth. As respects water management, it means working for the ecological protection of our river basins and waterways, not for full development.

We must adopt water management policies that will restore the pristine purity of our rivers, lakes, and estuaries. That means a new set of standards. But the weight of influence in Washington, D. C., is against them. The answer is, "Those goals are fine but we must seek to attain them only to the extent practicable."

What are those standards?

(1) Our waterways must be swimmable once more.
(2) Our waterways must no longer be used as a waste treatment system.
(3) They must protect fish, shellfish, and wildlife, and support man's life and health as well.
(4) The discharge of toxic pollutants into our waterways must come to an end.

Those are the ecological "musts."

RADIATION

Leukemia appears in radiation-exposed persons approximately five years after exposure, whereas most other cancers take 10 or more years to occur. Therefore, a group of humans exposed to ionizing radiation would show only leukemia five years later, simply because all of the other cancers had not yet occurred. The expert bodies of scientists studying radiation hazard for humans fell into this specific trap and as a result, they seriously underestimated the cancer hazard.

What is even worse, the error has been even further compounded. Knowing that some forms of cancer may take even 15 to 20 years to appear after radiation, these expert bodies still were refusing to consider additional cancers even though they realized it might still be too early, 15 years after radiation, to perceive the full effect.

JOHN GOFMAN & ARTHUR TAMPLIN,
Poisoned Power

Study of the Columbia River, on which the Hanford, Washington, reactor is located, revealed that while the radioactivity of the water was relatively insignificant: 1. the radioactivity of the river plankton was 2,000 times greater; 2. the radioactivity of the fish and ducks feeding on the plankton was 15,000 and 40,000 times greater, respectively; 3. the radioactivity of young swallows fed on insects caught by their parents near the river was 500,000 times greater; and 4. the radioactivity of the egg yolks of water birds was more than a million times greater.

RICHARD CURTIS AND ELIZABETH HOGAN
Natural History, MARCH 1969

We have concluded that there are major and critical gaps in present knowledge of safety systems designed to prevent or ameliorate major reactor accidents. We have further concluded that the scanty information available indicates that presently installed emergency core-cooling systems could well fail to prevent such a major accident. The scale of the possible consequent catastrophe is such that we cannot support the licensing and operation of any additional power reactors in the United States, irrespective of the benefits they would provide.

BY IAN A. FORBES, DANIEL F. FORD, HENRY KENDALL AND JAMES J. MACKENZIE
MEMBERS OF THE UNION OF
CONCERNED SCIENTISTS OF BOSTON,
14 ENVIRONMENT, JANUARY-FEBRUARY 1972

RADIATION IS EVER PRESENT. Atomic bomb testing has given the planetary winds radioactive particles to carry.

A nuclear plant exhales radiation, and AEC says that it is permissible for any one plant to emit 44,000 curies a day. Is that too great an exposure for people?

Radioactive waste material is another hazard. Where will it be buried? How will it be transported to its burial grounds?

Much has been buried in basalt hills in the state of Washington. But this basalt or lava is water-bearing, some wells producing water at 10,000 feet. A flow system in basalt commonly returns the water to the surface at some point on the hydraulic gradient. Some think that our Hanford, Washington, buried wastes will remain "hot" for tens of thousands of years, plaguing untold generations through leakage.

The tanks at Hanford, which hold up to a million gallons each, are made of concrete and are steel-lined, and they have an estimated life of twenty to thirty years. There have been twelve leaks in these storage tanks since 1958, with 300,000 gallons being lost. One leak in 1970 was 70,000 gallons. But the men at Hanford say the leakage never reached the water table. The water table has, however, been contaminated by radioactive wastes, as a result of putting some low-level wastes into ponds. AEC insists, however, that the resultant contamination is within the allowable limits.

But a report of the National Academy of Sciences to AEC, made in 1966, and not released except under the pressure of Senator Frank Church and his committee in 1970, condemned such disposal even of low-level wastes:

"The Committee thinks that the current practices of disposing of intermediate and low-level liquid wastes and all manner of solid wastes directly into the ground above or in the fresh-water zones, although momentarily safe, will lead in the long run to a serious fouling of man's environment."[1]

Senator Mike Gravel said in 1971:

"If as little as one per cent of our present annual production of radioactive garbage is lost to the environment, the equivalent of about 100 Hiroshima bombs would be contaminating this country every year. Can the AEC account for —locate—99 per cent of the radioactive garbage already produced in this country?"

Waste piles from uranium mines are washed by rains which leach out radioactive materials. This has dangerously polluted some waters in the Colorado River Basin.

Some uranium mine waste fills were used for construction of homes and schools in Colorado. This permitted radon gas, a product of radium in mine wastes which is heavily implicated in lung cancer, to permeate these buildings. These tailings were also used for construction in seven other Western states.

A single uranium mill in Grand Junction gave away about 200,000 tons of tailings in the fifties, which were used in over 11,000 structures. Abnormally high radioactive readings have been found in a dozen Colorado cities and towns as a result of the use of these and like tailings. As uranium decays it changes into thorium, which in turn changes into radium which turns into radon, and so on. The state officials are concerned but AEC says the disposal of tailings is none of its business.

One of AEC's plants—the one at Rocky Flats near Denver —had a disastrous fire that resulted in spreading dangerous plutonium into the surrounding area. Earth, dirt, and water became radioactive and the high winds infected nearby Denver and other population centers.

Tailings from uranium mines throughout the West and Southwest are radioactive and they have infected the fields and crops grown there, as well as waterways, including Lake Mead. Mills that process uranium ore in nine Western states have left, as of 1972, piles of radioactive wastes of about 100,000 tons. These tailings are ground smaller than sand and are blown away by the winds. In that way and by other means they find their way into rivers, polluting the water with radium that has halflife of 1,620 years. No plan for controlling or disposing of the tailings has been designed. That is why EPA in 1972 proposed a model law that would require the tailings to be spread out and contoured and then planted with grass, so that they might be stabilized and not become airborne pollutants. Those tailings will increase, for our commitment to the hazardous nuclear power means more and more digging, with the production of vast amounts of tailings.

One of the by-products of nuclear power plants is iodine-131, which the University of Nebraska has discovered now infects cattle thyroids throughout the West. Iodine-131 has a halflife of only eight days. But another iodine produced by nuclear reactors is I-129, which has a halflife of seventeen million years. That means, of course, that I-129 will remain in the environment forever, so far as human time is concerned. The alarming fact recently reported[2] is that an upstate New York nuclear fuel reprocessing plant is discharging most of its I-129 into the air and water. It has measurably affected flora and fauna. More reprocessing plants are planned. At the present rate of poisoning, radiation pollution will be permanent and irremedial. Seventeen million years from now, though we and our nuclear plants will have long since passed, I-129, which we created, will still be polluting the environment.

Nuclear plants around the world produce recurring accidents that pollute the countryside. Of the forty-two that happened in 1966, thirty-seven were in the United States; and in the same year six of our plants had more than one accident.

In 1957, Congress, after being advised that a single nuclear accident might cause $7 billion in damages, passed a law that limits public liability to $560 million per nuclear accident. The Act provides that the government will pay 83 per cent of that amount, the private utility only 17 per cent. In other words, the victims—the public at large—pay the bill.

Senator Gravel said in 1971:

"Every time we hear about the startup of one more 500-megawatt nuclear plant, it means that we are irrevocably committed to controlling and guarding about 250 megatons' worth of long-lived fission products for a few hundred years . . .

"In other words, every single nuclear powerplant inevitably creates a giant radioactive legacy. It runs [sic] equivalent to exploding about 1,000 Hiroshima-size bombs per year, per 1,000-megawatts of electrical capacity."[3]

Thermal pollution has been around for a long time. But the advent of nuclear power increases it mightily. A nuclear plant produces 60 per cent more heat than a conventional plant of the same magnitude. Dumping into the nearest river, bay, or lake is the easiest course. A one-million kilowatt plant, putting 3,000 cubic feet per second into a river, raises the temperature of the river about 10°F. Water above 75°–77°F. is lethal to trout. Only some southern species of fish can live in 93°F. water. Spawning for trout requires water about 55°F. A rise in the temperature of the Columbia River by 5.4° F. would be disastrous for the eggs of the salmon. Cooling towers are one alternative and cooling ponds another. Whatever method is used, the impact of nuclear heat on the environment is going to be appalling.

Nuclear power plants use an enormous amount of water. Nuclear energy fans call those ecologists who oppose thermal pollution *eco-kooks.*

We do not know all the dangers of thermal pollution. We do, however, know that it will eliminate many of the food items on which birds live. If thermal pollution eliminates 25 per cent or more of the diet of a bird, the bird itself is threatened. At some wildlife refuges, that means that over one-third of the birds will be on the endangered list.

Apart from thermal pollution, anxious citizens press hard for proof that nuclear reactors are safe, for proof that a prudent person can live downwind from a reactor without fear, and for proof that disposal of waste materials does not create an ongoing risk that may last 24,000 years. The ecologists say that by the year two thousand the number of six-ton tankers in transit with radioactive wastes will be over three thousand—"a mighty number of curies to be roaming around in a populated country." The opposition fears that the tendency to cut costs for the benefit of stockholders of these new private power companies will lead to relaxed standards and disastrous accidents. And so the lines are drawn as the 492 nuclear power plants projected for 1990 are designed.

As Senator Gravel said in 1971, "If this country ever grows dependent on nuclear energy for more than a few per cent of its electricity, the entire economy could be crippled by one bad nuclear accident which required the shutdown of all nuclear plants."

How safe is it to go ahead with nuclear reactors for peaceful uses? Once the ground, water, and air are dangerously polluted, there can be no return to safety because of the length of the halflife of most of these contaminants.

John Gofman of the University of California and other scientists plus some laymen who make up the Committee for Nuclear Responsibility, Inc., at 110 East 59th Street in New York City, say that there is "no solution in sight for the legacy of radioactive waste being generated now, and to be increasingly generated by this industry." That Committee says, "Unless *all steps* combined in the nuclear power industry can be carried through with 99.99% containment of radioactive wastes for centuries to come, the earth will be irreversibly polluted and degradation of humans and other species will be inevitable."

The official program is more dangerous, health-wise, than nuclear energy in its conventional form, which splits uranium atoms to produce heat. Only a small fraction—less than one per cent—of uranium as it is mined is suitable for nuclear fission. The "fast breeder" process converts uranium into plutonium, which serves directly as fuel. Thus the reserves of uranium are conserved. Whether the "fast breeder" can be made to work commercially is not known; the Fermi plant in Michigan has not been a success during twenty years of operation. A reactor core meltdown (which could trigger an atomic explosion) released radiation in the plant and put it out of operation for five years. It is, however, now operating at low power; and the American kite is now tied to the "fast breeder."

Late in 1971 the Argonne National Laboratory, near Chicago, announced that the "fast breeder" will be a major source of electric energy in another fifteen years. The manufacture of plutonium, however, carries the risk of pollution of the environment through catastrophic accident. Plutonium has a halflife of 24,400 years; thus the present official program casts a massive pall over generations yet unborn.

John Gofman and his Committee denounce "the breeder reactor," saying "it will commit humans to an energy economy based upon plutonium—the most toxic element known to man." The Committee adds this warning: "We will be handling hundreds of tons of plutonium per year; yet one *pound* of plutonium is enough to produce approximately 10 billion cases of human lung cancer. Further, any plutonium released upon the biosphere will remain radioactive for tens of thousands of years to produce human lung cancers in future generations. On top of all this, plutonium finding its way into underworld hands will make fabrication of nuclear weapons by private groups and non-nuclear nations into a probability."

The dangers are greater even than those mentioned. Chemicals already in our environment may increase the cancer-inducing capabilities of radiation. Some such sensitizers are known and others will be discovered. Much research needs to be done at this level. Meanwhile, placing nuclear facilities near urban areas, where a carcinogenic environment already exists, may be extremely dangerous.

The National Academy of Sciences in its 1966 Report to AEC said that since the waste material will be radioactive for such long periods, "neither perpetual care nor permanence of records can be relied upon."[4] Storage of liquids is unsafe; the radioactive wastes must be reduced to *solids* and stored in caverns of salt beds; and that is now being done in Kansas, near Lyons. But the enormous risk of the *liquid* wastes now stored by us, by Russia, England, France, and China, remains appalling.

The radiation experts are divided on whether the present radiation in the air is so great as to endanger man. AEC radiation guides allow 171 milligrams per year. Some radiation experts object, saying that exposure should be reduced to 17. The difference, they say, may be 30,000 cancer cases a year.

In New York some scientists, opposing the spread of nuclear plants, maintain that the presence of existing reactors has already increased infant mortality. Some are beginning to think that even man's germ plasm may be imperiled if we seek to live under the existing AEC radiation guidelines.

Among those who say AEC radiation guides should be reduced to one-tenth are John Gofman and Arthur Tamplin of the University of California. In March 1970, they made a further downward revision, concluding that the ceiling should be lowered to "absolute zero." They now say that nuclear energy for production of power is not needed. "If you look at what's happening, the only thing that power generation really correlates with in the United States is the production of garbage."

The National Radiation Council in 1971 denounced those critics. Gofman replied that the Council could not be believed because many of its members were government employees or dependent on the government for their living. And so the battle of the experts goes.

The National Environmental Policy Act of 1959 requires

AEC to present the "adverse environmental effects" of technology in the statements that it files. (In 1972 Congress gave AEC temporary exemption from that Act for previously filed permits.) But Dr. Ralph Lapp testified that the layman who challenges AEC needs "time, money, and availability of competent technical authority" in order "to match wits with the nuclear utility." To get that expertise, the layman needs to be funded, Dr. Lapp maintains, by the federal government; and without lay objections, there may not be any check "on development of unsafe reactors or on unsafe siting."

Tamplin and Gofman with their book *Population Control Through Nuclear Pollution*[5] and Curtis and Hogan with their book *Perils of the Peaceful Atom*[6] lead the way in arguing for delay in the nuclear energy field.

Some states are beginning to rebel at AEC domination and are trying to impose stricter radiation standards than does Washington, D.C. The states in the vanguard are Colorado, Minnesota, and Pennsylvania. Senator Gravel proposed in 1971 that there be a one-year investigation of the engineering and biological hazards with regard to nuclear power and that the investigation be followed by public debates between advocates and critics before unbiased scientific juries.

AEC has been having problems. Its "boiling-water" type of reactor is apparently more radioactive than its "pressured-water" type. The latter was originally designed for use in nuclear submarines but was more expensive. The former was cheaper to build and operate. Moreover, the cooling system of the reactors has come under severe attack, and some in Congress are concerned that AEC's Safety Research Division is either too lax or too industry-oriented.

Consumer interests demand that we rely on other sources for the development of energy. AEC by 1971 had told so many half-truths, had served as a Madison Avenue promoter rather than as a scientific watchdog for the public, that it was experiencing the same kind of credibility gap that forced Lyndon Johnson from public life in 1968.

The new head of AEC, James R. Schlesinger, seems to realize the existence of that credibility gap. In 1971 he was saying that AEC is not a front for industry but exists "primarily to perform as a referee serving the public interest."

As to other types of power, it is said that smoke from conventional fuels is harmful. And so it is, since it dispels mercury, among other pollutants. Mercury is in the air as well as in the waters. It comes from the burning of coal and of crude oil. Mercury is found in Asian rice, in American wheat and in a wide range of foods from baby foods to walnuts. Hair samples show mercury in generally high quantities. Hospitals may use fifty pounds or more of mercury a year to make mercuric chloride, which is used as a cleanser. All of this in due course reaches the environment. Smokestacks discharge sulphur, particulates, and other harmful chemicals. But we know enough to make it possible to reclaim those chemicals. That should be our course.

Some of our scientists and the Russian scientists claim that we are well on our way to harnessing hydrogen fusion. A hydrogen fusion plant has enormous advantages. (1) It can be built almost anywhere. (2) It would be "inherently impossible" for a fusion plant to have a nuclear accident. (3) A fusion plant would cause no air pollution. (4) And it would be practically free of radioactive wastes.

Keeve M. Siegel (former professor at the University of Michigan and now head of a scientific research firm) predicted in January 1972 that in five years hydrogen fusion would be feasible. A safe, compact, and pollution-free generating plant could be installed in the basement of an apartment building and provide electricity for forty blocks around.

Fusion would entail the problem of disposal of hot water and raise prospects of thermal pollution. With fusion, however, we need not despoil our air and waters with radiation. But in 1971 the Administration cut research funds for fusion and heavily increased the budget for nuclear fission, thus setting back perhaps for a decade or more the arrival of a truly "golden age" of power.

There are other alternative sources of power. The scientists say that solar energy, which is also a radiation-free alternative to nuclear energy, is now "technically feasible." Whether the economics of it can be resolved is the main question. The details are given by Wilson Clark in the *Smithsonian* for November 1971 and by Dr. William J. D. Escher in the *Congressional Record* for November 2, 1971. The plants are large-scale, movable, free-floating ocean-borne complexes. The products—liquid hydrogen and liquid oxygen—could be taken around the world in cryogenic tankers and further transported inland by vehicular or pipeline means. They could be converted to electrical power. This would be the "sunshine economy," not the "plutonium economy" toward which we seem to be headed. The "sunshine economy" would be pollution-free.

A recent study[7] by Norman C. Ford and Joseph W. Kane shows that solar energy can heat water so that some of the water vapor is thermally dissociated into hydrogen and oxygen, the hydrogen being siphoned off and used, as natural gas is used, to generate electricity. "It is pertinent to note," they say, "that only two per cent approximately of our major deserts are capable of producing enough hydrogen to equal all of the electrical power we now use."

On a much smaller scale than Dr. Escher proposes, thousands of solar water heaters are already in use in this nation. Some of our houses have solar space heaters and solar air-coolers. Russia had vast solar energy plants in Soviet Central Asia as early as 1955, when I saw them operating.

One aspect of solar power is sea-thermal power. Scientists tell us that there is enough heat in the Gulf Stream alone to supply two hundred times the total power requirements of the United States. The heat transferred from the surface water of the ocean will cause propanes in a boiler to evaporate into a high-pressure vapor which drives a turbine that in turn generates power. Cold water, drawn from the lower levels of the ocean, reduces the temperature of the propane in a condenser. The same propane is continuously circulated through the power cycle. Sea-thermal plants are safe and simple; they do not pollute the air and water with radioactivity, noxious gases, particulate matter, or odors.[8]

There is also the prospect of geothermal power, symbolic of which are the geysers prominent in a few of our national parks. In 1970 Congress enacted the Geothermal Steam Act of 1970,[9] which authorized Interior to lease public lands for geothermal development. There are those who see a rosy future for geothermal electric power. There are others who fear the environmental costs will be enormous. As this is written battle lines between the two are being drawn.

Wind power is another source of energy. Wind velocity maps show that many states, particularly in the Midwest, have winds averaging twelve miles per hour or more. Wind of course is variable; but studies show how a wind-power generator can decompose water into hydrogen and oxygen, which would be stored under pressure. A fuel cell would then recombine them and pollution-free generation of electricity on a steady basis would result.

Still another source of energy is the magneto-hydrodynamic power generator. Russia has the only one in the world, and it began operations in December 1971. An MHD plant does not use steam to turn a turbine. Instead, magnetically charged particles of gas or a liquid cut through the field of a magnet, generating voltage in the gas or liquid. This voltage is drawn off as current by electrodes. Research on this system is being conducted in this country, but no plant operates here. MHD is said to be particularly desirable because it can attain efficiencies up to 60 per cent, compared with 30 to 40 per cent by other processes. Russia plans a 1500-megawatt MHD facility by 1978 which, it is said, would put Russia five to ten years ahead of American technology.

Federal expenditures for research and development in the field of energy are about $500 million annually. Nuclear power gets 85 per cent of it. The rest goes to problems of fossil-fueled power facilities. NASA does research on solar energy as related to outer space use. Small sums have been advanced for MHD research. But neither solar energy nor sea-thermal power energy nor geothermal energy nor wind power have been subsidized for research and development.

The priority given nuclear energy as respects R & D funds has caused even fossil-fueled problems to multiply. The technology to remove sulphur oxides from the combustion process has not been developed. And the same largely holds true of nitrogen oxides. Neither federal nor state governments became seriously interested; and industry did nothing. Not until 1970 did Consolidated Edison in New York make any real effort to develop stack gas-removal processes,[10] although long before that time New York City was suffocating from sulphur oxides and nitrogen oxides. The failure to remove the sulphur is truly shameful. About 12 million tons of sulphur are emitted into the air each year, yet we use 16 million tons, all of which is mined, rather than retrieved.

The reasons for giving priority to nuclear power are not difficult to discern. The powers-that-be want to protect their investments in uranium stocks and in the millions that have gone into R & D on nuclear energy. Nuclear energy represents an investment of three billion in tax dollars and twenty billion in private dollars. As John Gofman and his Committee for Nuclear Responsibility say, "Millions of dollars can and will be spent annually to protect this investment from criticism or a nuclear power moratorium." Those investments would be lost in a regime which used hydrogen, available in the air, free, or solar or geothermal energy. Aside from hydrogen fusion, those who expect that rational decisions will be made on the choice of fuels are mistaken. The myth is that coal, gas, oil, water power, and uranium—the five major energy sources—will compete with one another to provide the best and safest energy sources at the lowest prices. But as Ridgeway, in *The Politics of Ecology,*[11] shows, the old trusts are moving in to corner the critical markets in these fields. Oil companies in the last five years have, indeed, purchased eight of the ten largest coal firms, which account for 50 per cent of our coal production. Of the nation's top fifty coal producers, twenty-nine are subsidiaries of oil companies.

The problem of energy demand comes back to the problem of "growth." Our officials say that by 2000 A.D. our annual energy demand will have tripled over current requirements. That raises the question of the kind of "growth" that is visualized. If anything goes, then we are in for a vast increase in the production of garbage, more land denuded by strip mining, more heavy metals in our fish, a mounting pollution of the waters, and an increase of solid wastes with no place to put them. An increase in production and consumption for their own sakes is not only wasteful but at times lethal, and always immoral in a world of want and hunger.

In Washington and Oregon, 25 per cent of the electrical power goes to operate aluminum plants; and their effluent has fouled our rivers and even the orchards watered by irrigation canals. Is the use of aluminum cans for beer the kind of growth we want? It is said that we must have more electricity because we are going to have more gadgets which will make us more comfortable. But as Wendell Berry says, "This, of course, is the reasoning of a man eating himself to death."

Before we go headlong for nuclear energy plants and for multiplying other sources of energy we need to determine priorities in this concept called "growth." The top priority should be the creation of enough energy to undo the damage that we are doing daily to the environment. Energy for recycling wastes; energy for mass transportation; energy for sewage treatment plants and for clearing up our water courses; energy for wiring the new 26 million units of housing the

Congress has ordered; energy for removing noxious gases and particles from our smokestacks and exhaust pipes of cars and trucks; and energy for the new research needed in pollution-free energy plants—costs which Simon Kuznets in *Economic Growth of Nations*[12] calls "diseconomies."

We need to place a stop-order on our "growth" for growth's sake. The standard for "growth" should be qualitative improvements in living. We can be happy without the new electrical appliances that flood the markets. We do not need the environmentally controlled office buildings that are multiplying. Those opposed to this position say that the environmentalists propose freezing the poor into their present inadequacies and keeping the underdeveloped nations down under. That is not a fair charge.

It is, as Barry Commoner recently said, "utter nonsense to say we have to double power capacity every 10 years."[13] Rationing is common in our economy. It pertains not only to commodities but to other facilities. Our city parks use permit systems to ration picnic grounds. The Forest Service, to prevent overcrowding and overuse of the John Muir Trail in the Sierras, requires permits for hikers, travelers, and campers. New York's "brownouts" in summer months, like the "brownouts" of other large cities, are due to the strain on power capacity because of the heavy use of air conditioners. Yet as Barry Commoner says, "By shutting down four cement plants and two aluminum plants upstate, you might eliminate the summer crunch."

Or as Professor Herman Daly of Louisiana State has said: "Central heating in the winter makes it too warm inside to enjoy the hearth fire, which man has loved from earliest times. In the summer air conditioning makes it too cold inside, and thus possible to enjoy a fire. Such self-cancelling uses are capable of absorbing great amounts of electric power, and perhaps of jading consumers' tastes to the very limit."[14]

On analysis, the crying "need" for additional power in the

Pacific Northwest comes down to quadruple aluminum production by 1987, the heating of more homes by electricity, and the trebling of the use of electrically driven appliances. Yet those who are *not* promoters of electrical processes or electrical equipment can show that we could lower our consumption of electrical energy below our 1968 levels without sacrificing our standard of living.

A policy that curbs "growth" implicates American ethics and philosophy. In an uncrowded world where there were new opportunities beyond existing frontiers, few restraints were palatable. But now we know that the earth has been occupied, that its inhabitants have increased, that we all live in a house that is big but cramped. The ethical system to which we were accustomed is inadequate for a crowded world. Actually we face a crisis so complex that none can fully comprehend it. We need social innovations as radical as our technological revolutions.

What "growth" that is dependent on additional energy do we need to improve the quality of life? That should be our national goal. A national energy policy will have to be articulated, argued, and debated. The ecologic ethic that was adequate for an uncrowded world is inadequate for the overcrowded world we now have. Several innovations, as startling as our technological capabilities, are needed to gear our demands to the finite, enclosed life-support system known as Earth.

Nuclear energy presents a world problem. While there are 55 new nuclear plants under construction here and 44 others planned, there are at present 286 power reactors in 22 nations. By 1980 there will be power reactors in 30 nations. It is estimated that by 2000 A.D. at the present rate, one new 1000-megawatt nuclear plant will be brought into action every day. The prospects of universal poisoning of the earth through peaceful use of the atom are now real.

The International Atomic Energy Agency of the UN is a potential watchdog. But some workable international controls are essential before anyone can truthfully say that the great potential benefits of nuclear energy are worth the awful risks that it entails. Yet local plans to promote it go ahead feverishly, as though no warnings of suicide had been given.

PESTICIDES

As crude a weapon as the cave man's club, the chemical barrage has been hurled against the fabric of life—a fabric on the one hand delicate and destructible, on the other miraculously tough and resilient, and capable of striking back in unexpected ways.

. . . in terms of the number of species, 70 to 80 per cent of the earth's creatures are insects. The vast majority of these insects are held in check by natural forces, without any intervention by man. If this were not so, it is doutbful that any conceivable volume of chemicals—or any other methods—could possibly keep down their populations.

RACHEL CARSON,
Silent Spring

It is now well known that pesticides can be dangerous to man and to wildlife too. Birds are disappearing—the brown pelican, peregrine falcon, and others. Schoolchildren know that DDT makes the eggshells too soft for nesting. Fish are also disappearing; and schoolchildren also know that DDT makes the female sterile.

Pesticides infect us and our food. Pesticides used on the land are eventually drawn into the ocean, and in recent years were found to have a lethal effect on sea life. The United States and India are the greatest users of DDT, and it drains to the oceans where it kills crabs and shrimp. Oysters store 70,000 times the amount of the DDT in the surrounding waters. Plankton, on which all forms of sea life are dependent, decreases in productivity by 50 per cent to 90 per cent when sea water has one part of DDT to one million parts of water.

The disappearance of commercial fish from the Atlantic is alarming. DDT is thought to be the culprit; and the fear persists that the same fate awaits the remaining commercial fish off New York.

The ocean off California is probably more infected by DDT than any other. Brown pelicans and cormorants have now failed to reproduce along those shores. The peregrine falcon is almost gone; murres and ashy petrels are in grave danger, as are blue herons and common egrets; bald eagles and ospreys have disappeared. Sea lions are on the decline. From California to Mexico all the seabirds and mammals that eat fish are disappearing. And the fish they eat "inhabit waters that are among the most pesticide-laden in the world."[1] The osprey hangs on precariously on Long Island, New York.[2]

Sterilization of fish is only one aspect of the tragedy. Studies show that some fish in the Darwinian selective fashion become immune to pesticides. For example, the tiny gambusia, the smallest of all fish in our southern waters, has at times become so immune to dieldrin as to become transformed into a living lethal sac of poison which kills anything that swallows it. The ramifications of pesticides in rivers, estuaries, and oceans are vast and as yet not wholly known. But the specter of the fish which survive becoming potent carriers of poison should alarm laymen as well as specialists.

Our Western yellowjacket is a valuable biological control agent, feeding on flies and caterpillars. Pesticides cover a wide spectrum, killing other control agents as well as yellowjackets. Attractants have now been developed that are used in catch traps placed around picnic areas, orchards, and beehives to catch the offending yellowjacket, yet not to wipe him out.

The Forest Service uses herbicides extensively in its so-called brush control program. Herbicides administered from the air drift off target; and at least 50 per cent of the amount of chemicals sprayed reaches adjoining areas that contain trees. Herbicides affect trees, as our experience in Vietnam shows. And there is evidence that forests which have been sprayed are likely to be more flammable. The Forest Service has been so loose with its spraying projects[3] that some have suggested for its motto, "Only We Can Prevent Forests."

In the Southwest we spray mesquite trees, wiping out in the process a long chain of life on which owls, coyotes, and other predators live.

In the Rockies we spray to eliminate sagebrush and in the process kill willow as well. Willow is the sustenance of moose and of beavers too and its disappearance has brought an end to these animals in the sprayed area. Nothing was achieved in terms of grass improvement that could not have been accomplished by restricted grazing. Why should beaver and moose be made to step back for man so that he can produce a bit more beef that we do not need?

Who authorizes the spraying? A faceless, nameless person in the Forest Service. His is the sole authority. We the people are not even allowed a hearing to offer proof of what spraying will do to the ecosystem.

Or the local park authority hires an outside sprayer who

selects the pesticide to be used. That tragedy was illustrated in Moscow, Idaho, where the contractor chose Bidrin to kill aphids, scale, elm leaf beetle, and red spider mites. The advantage of Bidrin, it is said, is that it "gets them all"—including the birds.

Our TVA—struggling hard to become Environmental Public Enemy Number One among the federal agencies—uses the herbicide 2, 4-D (which causes human birth defects) to kill off milfoil, which grows in the artificial lakes which its impoundments have created. It spent one and a half million dollars on 2, 4-D between 1962 and 1969 and sprayed it on 35,000 acres of its reservoirs.

Michael Frome reported in *Field & Stream*, March 1972, that the Park Service "celebrated Earth Week, a period of ecological commitment last year, by spreading methoxychlor, a hard pesticide cousin of DDT, in at least one area along the Potomac."

The effect of pesticides on predatory birds is greater than it is on songbirds because: (1) they eat food in which pesticide residues are very large, and (2) their reproduction rate is so low that they have no chance of developing resistant strains.

Countless scavenging organisms such as burying beetles have to run the gauntlet of soil pesticides. Freshening bacteria and fungi of the soil are discouraged with fumigants. Maggots often cannot play their salvaging role because the flesh of intended victims contains toxic levels of pesticides.

The use of melathion at Lake Tahoe for mosquito control apparently wiped out the tiny wasp parasites that normally keep the pine needle scale under control.

When pesticides make the target pest resistant to them, they become more and more inefficient. As a result the amount used is increased, and all is well with the chemical companies. Thus in the last twenty years the amount of DDT per unit of produce increased by 168 per cent. The decreased efficiency means an increased impact of the insecticide on the environment.

Now we have mosquitoes that are immune from all chemical controls. One is *culex tarsalis,* the one that carries sleeping sickness.

There are many aspects to the pesticide problem. Oklahoma is losing 100,000 bobwhite a year. They cannot survive without the supercharged proteins that insects alone can provide. Pesticides tend to eliminate an important insect at a critical time of the year.

Spraying with herbicides to kill marijuana, the Indian hemp plant that grows over an eleven-state area in the Midwest, will, it is feared, hit the pheasant population hard. For the herbicides promise to take away nesting cover and protection of many plants that are needed in the winter against the elements. Moreover, the hemp seed is a preferred food of quail, doves, pheasants, and songbirds. What we do in this crusade by the Bureau of Narcotics has vast ramifications in the world where our birds live.

There is a pesticide called Sevin, used by corn growers, which, it has been discovered, wipes out entire bee colonies. The fruit growers of Oregon and Washington need bees for pollination. As a result, some two billion honeybees had to be airlifted into the Pacific Northwest in the spring of 1971. In Arizona the bee kills have hurt the production of long staple cotton and wiped out melon production in the Salt River Valley.

Sevin is also used on the gypsy moth although a biological control, which does not affect its natural enemies, has at last been found. Every seven years it shakes off its parasites and predators and defoliates whole stretches of forests. Some individual trees—but never the forest—die; the main stand quickly recovers. But the chemical companies spread the alarm and even groups like the Park Service start spraying with Sevin. Sevin, however, inhibits the growth of mammals and birds which control the gypsy moth. The result is that Sevin merely thins out the gypsy moth, making survival possible for more moths than would otherwise be possible.

Thus the chemical merchant makes more money simply because his product does not work.

Now the PCBs (polychlorinated biphenyls), which are odorless and colorless and used in over two hundred chemical compounds, are appearing in men, birds, and fish around the world. Isolated incidents of high-level contamination, resulting in death, are increasing.

The list of poisons and dangerous practices is long and depressing. Pesticides are not the sole culprit. Wildlife flourishes in a varied habitat of shrubs, forbs, and grasses. Now we farm the land clean, no longer tolerating the sunflower, thistle, and pigeon grass. By clean farming we create an ideal environment for a particular insect that specializes in a particular crop. Wildlife flourishes where there is variety in the habitat, not in a single-purpose environment. As a result of the creation of a single-purpose environment, South Dakota's pheasant population has dropped from 13 million in 1940 to 2 million at present.

Mercury poisoning has become a new terror. A mercury compound has been used to treat seeds, and it infects birds that eat the seeds.

Herbicide 2, 4, 5-T, the one used in Vietnam for defoliation, is used by the Forest Service to kill chaparral, in the interests of cattlemen. But the poison also causes deformed animals—mice, rats, goats, ducks. There is dispute over whether deformed babies are being born in Vietnam.

Agent Orange is the tough military version of 2, 4-D and 2, 4, 5-T used widely by ranchers to kill weeds and bush. We used over one and a half million gallons of Agent Orange in Vietnam and have 800,000 gallons in storage here. We used Agent Orange in Vietnam to kill crops and to ruin timber resources, particularly mangrove forests. We have indeed defoliated one-seventh of the land there, so that we will forever be remembered as the Great Despoiler, taking the place in history of Genghis Khan. Though we announced that Agent Orange would not be used in Vietnam, we have two deadly substitutes which we pour and spray over Vietnam: Agent White and Agent Blue. The despoliation has been so great that Senator Gaylor Nelson believes that "South Vietnam would have been better off losing to Hanoi than winning with us."

The surplus of Agent Orange is being sought by ranchers as government surplus. EPA is worried about it and has tried to reduce its use around home, water areas, and on food crops. But every attempt to limit its use has met with challenges in court. The future of Agent Orange is curiously in doubt, though its known damage is alarming.

The Department of Agriculture has been more or less the refuge of the chemical companies, as the Fountain Committee of the House revealed. Enforcement (until recently transferred to EPA) was in the Pesticides Regulation Division (PRD) of Agriculture. It was supposed to work in cooperation with HEW in determining the possible effects on human health.

The November 13, 1969, Report of the Fountain Committee of the House revealed that during the five-year period ending June 30, 1969, HEW objected to the registration of 1,600 pesticide products. Yet PRD registered "many, if not most" of those products.

When the federal agency issues cancellation notices for a pesticide the sale of it is not halted. The manufacturer may appeal and the procedure is long, drawn-out and time-consuming. It starts with an administrative panel selected from a list of the National Academy of Sciences, a public hearing, and appeals through the courts.

If the agency finds that a pesticide is "an imminent hazard to the public," it may suspend it, in which event interstate sales stop at once. But this seldom happened.

Lax enforcement by PRD was the charge of the Fountain Committee. Though proof of the safety of the product was on the applicant, PRD shifted the burden by registering poisons over objection, unless the objector could produce satisfactory evidence of hazard. Free enterprise still carries the day.

The new Environmental Protection Agency, created in 1970, now exercises the numerous controls of pesticide chemicals formerly entrusted to Agriculture, Interior, and HEW. Will chemical-company influence and control follow those transfers to EPA?

EPA is headed the other way. It was proposing, early in 1972, rules which would give any individual or organization the right to ask EPA for an immediate suspension of a pesticide that was shown to be an imminent hazard to health. And, in less critical cases, a like procedure was proposed for cancellation of a registered pesticide on a slower timetable. Public participation—not hidden chemical company control—seems to be the trend.

In 1971 EPA canceled the herbicide 2, 4, 5-T for use on food crops for human consumption; and that case has gone into hearings. EPA limited the use of DDT; but after a lengthy hearing the examiner said DDT poses no dangers to humans and does not have a deleterious effect on fish and wildlife. EPA at once took an appeal. In 1970 EPA suspended alkyl mercury pesticides to treat seed. In 1971 EPA canceled three mercury pesticides used to kill algae in swimming pools and water-cooling towers. EPA in March 1972 suspended federal registration for all alkyl mercury pesticides and all other mercury products applied to laundry fabrics and rice seed, and mixed in marine anti-fouling paints for vessels.

Then in 1972 EPA was proposing rules that give the chemical company the right to immediate administrative review of the suspension decision, a step which was paralleled in Congress by giving the chemical company immediate review in the courts, thus changing the rule of the Nor-Am case in 435 F (2d) 1151. The chemical lobby's voice is loud.

Pesticides have served a useful purpose around the world in controlling diseases and in increasing agricultural production. They cannot arbitrarily be banned; but severe controls are needed.

The President's Science Advisory Committee reported in 1963 that one out of twenty acres in the United States is polluted by pesticides as a result of our domestic use. Other pesticides reach here—and other nations as well—through planetary winds and through the so-called acid rains. When water containing DDT evaporates, the DDT goes into the air with the water. It is thus carried around the world by the wind in a few weeks, even though its use was local and restricted to an isolated area not draining into an ocean. Sweden has about 3,300 tons of DDT in its soil, an amount more than twice the quantity used in Sweden in the last twenty years. There are few, if any, organisms without DDT left on earth, since DDT has entered all animal systems and contaminated all offspring even before birth. The saturation of the world's environment with DDT and its persistence means that the ingestion of it would continue at least twenty-five years even if its use were immediately halted. The carbon molecule is quite indestructible.

Developing nations may have a continuing need for DDT. But we can end our dependency on it or use substitutes. We can at least save our waters, birds, fish, and humans from its poisons pending the day when we get a United Nations Convention outlawing DDT just as slavery was outlawed.

A debate has been raging over the effects of pesticides. Defending them is Norman Borlaugh, outstanding agronomist, and challenging his thesis is Barry Commoner, distinguished biologist. Borlaugh says that DDT has done much to help eradicate malaria; and so it has, in relatively dry areas. But in most areas experience shows that in a decade or so the mosquitoes have become resistant to the chemical. Borlaugh maintains that elimination of pesticides for agricultural purposes would have a major effect on U.S. food production. Commoner replies that about an 80 per cent reduction in insecticide use could be achieved with no reduction in agricultural output, *if* the harvested acreage were increased by 12 per cent, which is the amount of land diverted from agricultural production by various official land-retirement programs. And so the issue was joined by the experts, the chemical companies keeping discreetly in the background.

A like debate goes on in Russia, where overuse and misuse of chemicals seem to be as common as in this country. There are no chemical companies and their profits to denounce. The Politburo, however, speaking through Premier Kosygin early in 1971, urged an increase in herbicides and pesticides and a 60 per cent enlargement of facilities for their manufacture. But there are rumblings among the lesser lights.

Up to now the profit motive of chemical companies in the United States has been the driving force. From here on the ecological impact of a pesticide should be known before it is marketed. Once ecological rather than dollar standards become our guide, no pesticide should be marketed unless its side effects on the ecosystem are known. That means creating control plots where the impact of the new poison on the entire chain of life in the soil and in the waters flowing from the soil will be known before the poison is turned loose on men, on flora, on fauna and on all living things in the estuaries and oceans.

As we enter 1972, legislation is pending for pre-testing of pesticides. The lines are tightly drawn. The chemical lobby naturally wants its free-enterprise system free from the quagmire of Washington, D.C., bureaucracy.

But the health and well-being of people come before profits. Or do they?

Rachel Carson, author of *The Silent Spring,* thought that the use of biological controls, rather than pesticides, was the answer.

In Congress Gilbert Gude of Maryland leads the effort to establish a pilot project that will develop biological controls for agricultural and forest pests, eliminating once and for all our heavy reliance on highly toxic and persistent chemical pesticides.

The issue is not related to pesticides alone.

The chemical industry introduces from three hundred to five hundred new chemical compounds a year, and a drive is on to submit them to pre-testing. Here again pre-testing for safety and environmental impact is being urged on the one hand and resisted on the other. The battle rages not only over the type of pre-testing to be required, but whether the time granted EPA to prevent manufacture will be short or long (90 days or 180 days).

Without adequate controls of the ever-increasing chemicals, Rachel Carson's dismal prophecy in *The Silent Spring*[4] will turn out to be the century's greatest understatement.

GARBAGE

Amidst the tended flames of this inferno I approached one of the grimy attendants who was forking over the rubbish. In the background, other shadows, official and unofficial, were similarly engaged. For a moment I had the insubstantial feeling that must exist on the borders of hell, where everything, wavering among heat waves, is transported to another dimension. One could imagine ragged and distorted souls grubbed over by scavengers for what might usefully survive.

LOREN EISELEY,
The Unexpected Universe

. . . In Garhwal, I found no red, green, or black shirts, no flags or emblems, no mechanisms, no motorcars or airplanes, but I did find a happy and contented people. I think the attitude of Himalayan peoples to western progress is best summed up in the words of a Tibetan, and Tibetans consider themselves superior to Europeans in spiritual culture. He said: 'We do not want your civilization in

Tibet, for wherever it is established it brings unhappiness and war.' It is a terrible indictment but it is true.

FRANK S. SMYTHE,
The Valley of Flowers

Because wealth is energy, people everywhere will be rich in power when integrated world-wide industrial networks are established.

A world-wide electrical energy network linking the day and night hemispheres would result in staggering economic gains. As Vladivostok sleeps, Con Edisonovitch's current would be channeled to California.

Industry works best as a world system. Newly emerging nations must realize that their independence depends on their participation in world industrialization.

BUCKMINSTER FULLER,
I Seem to Be a Verb

GARBAGE PILES HIGH IN THE CITIES; and even the campgrounds, parks, waterways and remote trails—even the tops of our highest mountains—carry the marks of litter.

We have made little progress with our refuse disposal. Over two thousand years ago, one who lived in Jerusalem took his bucket of waste to the courtyard where someone put it on a camel which transported it to the desert or fields. Today a contractor hired by the city collects and dumps our bags of litter into a packer—the device that has taken the place of the camel. The packer often drives his load thirty miles to a dump where it is buried.

Burning of garbage is often used instead of burial. Yet burning is so inefficient that one can often read a burned newspaper from an incinerator easier than he can read it on a newsstand. Most burning, however, is open-dump burning, which means we have not advanced very far since the days of Christ, so far as garbage disposal is concerned.

The problem of using garbage for landfills raises large questions. Pennsylvania's plan to bale Philadelphia's garbage and put it in deep strip-mining pits for burial had insidious aspects. For no one knows what poisonous leachates it might produce a decade or two hence, ruining the percolating waters. Hence the legislature outlawed it, except under special permits.

What is the remedy? To make housewives give up the use of these new disposables? To make it illegal to flood the streams of commerce with them? To put stringent controls on technology? Those controls are critical. But, as we shall see, reuse or recycling is equally critical.

Romantic ideas are often advanced for the technological solution of many environmental problems. In the Muskie hearings in 1970, New York City's plight and technology's answer were used as an illustration.

New York City uses over a billion gallons of water a day. It discharges about 8 million tons of solid waste per year, and it uses about 700 megawatts of electricity per year. It is argued that by the utilization of existing nuclear technology, a 12,000-megawatt power center could be built in New York City to process solid waste, which in turn could produce the entire electrical requirements for Manhattan Island. There would be enough heat left over to melt the ice and snow from the principal thoroughfares of New York City. This argument does not consider the dangers of such a plant in New York.

Technology has made the shredding of used cars feasible. About an eighth of the bulk of a discarded car is made up of glass, wool, leather, fabrics, tires and other combustibles. While the metallic parts can be shredded or turned into pellets by grinding machinery, special incinerators are needed for disposal of these combustibles. About 20 per cent of America's junked cars are in Appalachia, a region that represents only 9 per cent of our people. Only a mass removal, the experts say, will do the job—either community efforts or a government bounty system. A bounty system for junked cars is better than a bounty system for coyotes, wolves, cougars, and bears.

We produce some 80 billion tin cans a year. There is some recycling of them, but in 1971 less than one per cent were recycled. But the steel companies have learned how to eliminate the tin from the can or how to use tin in ferrous scrap. It now is likely that all steel containers in the future will contain a minimum of 25 per cent recycled material.

Technology makes re-use of paper and paper products easy, and the economics are not forbidding. Yet more than 80 per cent of the production goes into one-time use. If in the next decade recycling is increased to 50 per cent, over 90 million acres of forest lands would be released for other purposes.

Recycling of paper wastes has been accelerated by the decision of General Services Administration, that does federal purchasing of paper, to obtain half of its stock from reclaimed paper. One-half means over $30 million in budget expenditures by GSA. In Maryland, recycled paper is now converted into fireproof home insulation and has become a booming business.

Throwaway bottles are more profitable to retailers and manufacturers. Yet throwaways use over three times the energy of returnable bottles, and cans use 2.7 times the energy of returnables.

We have a growing need for energy. At the same time we are pouring untold units of heat into our rivers and polluting them. How can this heat be turned into energy?

One pilot project located at San Mateo, California, burns solid waste and with the heat produced generates about 20 per cent of the power needs of the community. The Construction Equipment Associates, Inc., has more ambitious plans for Tonowanda, New York, and Brockton, Massachusetts; and Process Plants Corp. is already under way at Whitman, Massachusetts.

Fly ash contains microscopic glass bubbles called cenospheres, useful in making a plasticlike product for the manufacture of tables and chairs and for use in building deep-diving submarines. These sophisticated uses substitute a low-cost raw material for a high-priced one when manufactured.

Fly ash is now also used to make bricks—two-thirds the weight of ordinary brick and less than a third of the cost. Fly ash is used extensively in Europe for building materials. What are the costs of displacing brick kilns with those using coal ash?

A plastic bottle developed in Sweden turns to dust as a result of ultraviolet rays of the sun. How many vested interests would we have to unhorse to get a bottle like that to replace our glass bottles, our tin cans, our *deluxe* aluminum cans?

Some states and localities have banned throwaway bottles, Vermont leading the way. The trend will probably be to glass bottles since studies show that the average number of trips made by returnable soft-drink bottles is fourteen and by returnable beer bottles, twenty.

We mine lead as well as sulphur. Yet our air is polluted with them. We must recapture and recycle these substances!

We mine phosphates in some estuarine areas and destroy the estuaries in doing so. Yet phosphates are the chief pollutant from our best sewage disposal plants. We can reclaim them!

Our copper smelters poison our air, as we shall see. We need not mine much copper; we can recycle it.

Gold, silver, platinum are not thrown away but retained and reduced to other forms. Mercury has been thoughtlessly discarded. In the future it can be retained and probably reused. But the problem is to reclaim the tons that lie in the mud at the bottom of our waterways.

The problems are in large part technological. But reliance on the technological fix may be defeating. For the heart of all problems dealing with the environment is human attitudes, human desires, and human habits.

Technology has now developed a process for pulp mills that removes the substances that cause odor from the emissions of pulp and paper mills.

Detergents have been condemned for their phosphates, which speed up the eutrophication of waters and the resultant death of ponds and lakes. But they also contain emulsifying or foaming agents which do not break down in water. Suffolk County in New York on March 1, 1971, accordingly banned the sale of all laundry detergents because Long Island is largely dependent on ground waters, which are easily polluted by chemicals.

Congressman Reuss, who investigated the detergent companies from top to bottom, said that the federal bureaucracy had in effect become one of their departments. Hammering of committees of Congress on the soap industry continued, phosphates were largely driven out, and a new chemical called HTA took their place, until it became suspect as causing birth defects. The basic issue was whether profits earned for protecting a housewife's hands against redness were more important than protection of our waters against decay or the oncoming generation against birth defects.

Soap is made from oil and alkali; but once it enters the sewage system it decays. Bacterial enzymes break it down; and as a result only carbon dioxide and water return to the ecosystem. Thus the circle is complete.

Detergents, however, are different. Chlorine is needed in their manufacture, and mercury is used in the process, and in the end is released into the environment as a pollutant. Thus the circle is not complete. Moreover, the bacterial enzymes cannot break down the molecules of detergents which produce in streams and lakes the foam that eventually is processed by human kidneys. Under pressure bio-degradable detergents were developed; but they too had a catch. Benzene from them when discharged into waters was converted into carbolic acids quite dangerous to fish. Thus even the biodegradable detergents do not complete the circle. Caustics were another additive. The problem with the additives is that they do not complete the environmental circle. And phosphates, as already mentioned, literally kill our water with algae. Phosphates were sometimes used to soften hard water when soap was used. There are, however, other water-softeners.

Soap keeps us as clean as do detergents and does not foul the environment. All detergents should therefore be banned.

Buckminster Fuller says in *I Seem to Be a Verb*:[1]

"The Einsteinian dynamic norm replaced the Newtonian static norm and set in motion modern science and technology.

"Economists, politicians, and speculators believe that metallic resources are found only in mines. They do not seem to realize that mined metals may be melted and re-used. Every twenty-two and a half years, the world's used metals are revitalized and recirculated, and their quantity doubles.

Melt down all U. S. planes and you would have a larger tin supply than does the leading tin miner, Bolivia."

He added: "It will be cheaper to capture and re-use the chemicals being spewed from industrial smokestacks than it will be to redress the abuses of air pollution."

By 1971 that prophecy became a fact. The urban trash and garbage now collected and burned or buried is what the U. S. Bureau of Mines calls "urban ore." It can all be reclaimed and put back into the stockpile of raw materials, as shown by the operations of the federal Metallurgy Research Center at Edmonton, Maryland. The cost is $3.52 a ton; the end products are worth $12 a ton.

The Bureau of Mines has, at Franklin, Ohio, a pilot project that reclaims waste materials before incineration, a project which, it says, may be even cheaper than its residue system.

Collection of items of garbage for recycling has caught the imagination of many people. But it has not become a major institutional effort, in spite of the fact that of the fifty largest cities, about thirty have sufficient landfill space for less than ten years and twenty-five of them have sufficient space for less than five years. Cities with the least landfill space show the most interest in recycling systems. The prophecy therefore is that municipalities will have a mounting interest in the problem very shortly.

Some industries have now made recycling a public relations gimmick. A recent advertisement tells the story:

The American standard of living is
creating a garbage pile that is now a
national problem.

Garbage in.	*Valuable raw materials out.*
Non-Ferrous Metals	*Aluminum* *Copper* *Zinc*
Ferrous Metals	*New Iron & Steel products* *Scrap for use in copper industry*

Food Waste, Sewage, Sludge	*Soil Conditioners Organic Fertilizer*
Contaminated Paper, Plastics	*Fuel for generating steam & electricity*
Pure Paper	*New Paper Products*
Glass	*Paving Material Bricks, Block, Structural Shapes*

High freight rates for scrap are notorious. EPA is now studying (1) reduced freight rates for shipping scrap or recycled products; (2) more rapid amortization for purchase of such products; (3) government purchases of recycled products, and (4) a tax credit on recyclable items. The junk dealer is getting respect he never knew before.

Recycling, however, has problems that it inherits from prior phases of the life of a product. PCB (polychlorinated biphenyls) is widely used industrially. It makes sealants less brittle; it is a useful coolant; it is used in auto tires, brake linings, paints, and pesticides. It is also used in ink.

Recycled paper that was marked with that ink has PCB in it. Cardboard boxes used for articles of food commonly have PCB in them. PCB, which contaminates turkeys, chickens, fish, and eggs, is dangerous. If its quantity exceeds the federal standard of five parts per million, it is necessary to destroy the food, for PCB has killed people and caused genetic disturbances in animals.

The point is that recycling at times recycles dangerous poisons. That is why there must be an outright ban of PCB and other potentially harmful substances.

NOISE

It would have taken the sharpest of ears to catch the sound of a hermit crab dragging his shell house along the beach just above the water line: the elfin shuffle of his feet on the sand, the sharp grit as he dragged his own shell across another; or to have discerned the spattering tinkle of the tiny droplets that fell when a shrimp, being pursued by a school of fish, leaped clear of the water. But these were the unheard voices of the island night, of the water and the water's edge.

RACHEL CARSON,
Under the Sea-Wind

"Surely you must know that the stars are great hunters? Can't you hear them? Do listen to what they are crying! Come on! More! You are not so deaf that you cannot hear them."

I have slept out under the stars in Africa for too many years not to know that they sound and resound in the sky.

I hastened to say, "Yes, Dabe, of course I hear them!" But then I was forced to add, "Only I do not know what they are saying. Do you know?"

. . . he said, "They are very busy hunting tonight and all I hear are their hunting cries: 'Tssik!" and 'tsa!' "

LAURENS VAN DER POST,
The Heart of the Hunter

WE THINK OF NOISE as a local problem; and so it is. Yet the SST aircraft was a federal project; and the noise of jets is under FAA, a federal agency.

Noise is an increasing scourge of urban life and of country life as well. A decibel (db.) is the lowest sound that can be detected by the human ear. Loudness increases ten times with each 10-db. increase. Thus a level of 100 dbs. is ten billion times as intense for the eardrum as one db. Annoyance apparently extends from 60 dbs. to 90 dbs. Discomfort arrives at 110 dbs. Pain comes at 129 dbs. Permanent hearing loss at 150 dbs.

Near the beginning of the Three Hundred Year War, the man who disliked the noise of the city—the clop of horses' hooves and carriage wheels, the ring of the blacksmith's anvil, the hammer of the carpenter, the bustle of people and neighbors, even the clang of bells—could always go West to escape it. But one who seeks quiet today, whether in the solitude of a national forest or in the deepest crevices of the Grand Canyon, accessible only by raft, is as likely as not to find the silence shattered by the sonic booms of several jets. And for the urban dweller who must stay in the city, the problem is much worse: industrial noises assault him at work, traffic noises impinge on him coming and going, and he probably cannot sleep at night without the roar of many motors—jets and otherwise.

The noise level of the inner city runs about 85 dbs. If houses were twenty-five feet back from the street and if shrubs lined the avenue, the noise level would be reduced to 75 dbs. The poor with open windows are exposed to ten times the noise intensity of the air-conditioned affluent person.

Noise and the increasing pressure of population create tensions among people and among animals as well. Ear specialists report that natives of the Congo or Sudan can perceive frequencies as high as 28,000 c/s (cycles per second). Middle-aged people on Long Island, New York, an area infested with aviation noise, can barely perceive 8,000 c/s; a seasoned air pilot, 4,000 c/s. It is predicted that man in urban areas, exposed to mounting noise, may in time be born deaf. Scientists can show in detail how noise leads to deficiency and failure of cellular function. It is indeed reported that in the third century B.C. an emperor decreed:

"Criminals should not be hanged, but flutes, drums, and the chime of bells should be sounded without let up, until they drop dead, because this is the most agonizing death man can conceive. Ring, ring the bells without interruption until the criminals first turn insane and then die."

The sound of a jet taking off is 130 dbs.; and the greatest reduction technology can now promise is 10 per cent. Under a 1968 Act, FAA is given power to control and abate "aircraft noise and sonic boom." Some feel that FAA has such close ties to the aircraft industry that it is unlikely to impose effective controls. FAA allows planes to fly at a thousand feet over any "congested" area and five hundred feet over open and country places. But five hundred feet is perilously close to the eardrums and constant repetition will drive people "up the wall."

Near busy airports, such as Kennedy in New York, there are so many interruptions with the teaching process in schools as to make the "jet pauses" in a classroom consume about an hour a day. Only the production of "quiet" engines (which are within the power of FAA to require) can make living in those areas tolerable. There are pending bills which would not only require FAA to control and abate aircraft noise and sonic boom but also to give EPA supervision over FAA in that regard.

Overall, EPA has estimated that forty million people in this country are exposed to potentially hazardous noise.

The truth is at present there is practically no control of urban noise. Municipalities may have noise ordinances but they are seldom enforced. In 1970, however, Congress enacted the Occupational Safety and Health Act governing noise in all factories engaged in interstate commerce. The

Secretary of Labor has power to regulate noise by a code. The present code requires that the loudest continuous noise a workman may be exposed to during an eight-hour day is 90 decibels. Complaints are mounting that many factories cannot meet the standard. In opposition there are unions demanding that the noise level be further lowered. Meanwhile some manufacturers of quiet equipment are drooling over the sales prospects of their new offerings.

The federal government also regulates the noise level in factories manufacturing or furnishing supplies to any agency of the United States. A noise level of 90 decibels for an eight-hour day is the one provided under this law; and if "feasible administrative or engineering controls" fail to reduce the level, then "personal protective equipment" must be given the employees.

As this book goes to press, the House has passed a Noise Control bill and sent it to the Senate. The measure would give EPA power, *inter alia,* to establish noise-emission standards for new products and stiff civil penalties for violations.

The sonic boom creates a pressure of 1.3 pounds per square foot. That is enough to make some sandcliffs crumble in a wilderness area. In the rugged Cascades where we ride horseback the risks are different. One sonic boom will make a gentle horse jump several feet sideways. There is no problem when one is on even ground. But the risks are awful when horse and rider are on a narrow trail with a 500-to-1,500-foot drop-off on one side. The boom of the SST would put the sonic boom to shame. It would produce from 2.5 to 3.5 pounds pressure per square foot. There would be no possibility of controlling a horse under that shattering sound.

But that is irrelevant, the promoters said, for the SST would produce 150,000 jobs and generate up to 10 billion dollars in tax revenues. After all, we're for "progress," aren't we?

Happily, the Congress turned the project down.

EPA is now seeking legislation which would give the administrator power to limit "the noise-generation characteristics" of (1) construction equipment; (2) transportation equipment; and (3) other equipment powered by internal combustion engines (other than civilian or military aircraft, weapons, and the like).

There are over one million motorcycles, 700,000 pickup trucks, 600,000 four-wheel drive cars, 80,000 snowmobiles, and 50,000 dune buggies roaring and churning their way on public lands of the Bureau of Land Management alone. How high this figure would be if the national forests and national parks were included is not known.

Hundreds of thousands of dune buggies regularly assault the deserts and beaches of California. But they are also in Cape Cod, North Carolina, Maryland, Florida, the Rocky Mountain states, and the Southwest. Near the Mexican border up to 30,000 dune buggies race over a fifty-five-mile stretch of dunes. The buggies are as closely packed as swimmers at Coney Island. The bite of the tires is severe; and their destruction of the slight vegetation is enormous. The vegetation lives on borrowed time.

These motorized visitors are mostly marauders who are in the wild areas not because they love nature but because these unpatrolled areas give them a place to rampage beyond the reach of the law. Vandalism, misdemeanors, and major crimes are spawned by the adventurers, and the silence of the woods is shattered and the wilderness is pillaged.

As of 1971 there were nearly 2 million snowmobiles loose in this country. Some predict that once they become safe to operate, users will look for something more exciting. The popularity of this hazardous device is said to be due to the fact that people court danger that their dull jobs cannot provide. Some, indeed, say that prior to snowmobiles their winter days were dull and boring, which of course speaks poorly of American *homo sapiens.* The world of challenging ideas was always more exciting than whirling dervishes. But modern man, perhaps, is on some kind of a suicide cycle. At forty miles an hour, he leaves behind a trail of terrorized animals, destroyed vegetation, damaged farmland, and many irate and annoyed people. Snowmobiles have largely replaced dog teams among the Eskimos in the western Canadian arctic, and probably mark the demise of an ancient and vital civilization.

Snowmobiles have friends and foes. Friends point out how the advent of snowmobiles has brought winter business and prosperity to the Adirondacks. The foes point out the ever-present noise of the foul machines and their use of private property to go round and round and drive anyone seeking a sanctuary into near-insanity. The time has come for severe regulations.

Vermont leads the way on snowmobiles, setting a limit of 82 decibels at fifty feet for their noise level. Vermont also requires that operators of snowmobiles must get written permission from landowners before traveling on private lands.

The quietness of our wilderness is also being ruined by tote goats that roar into sacred precincts, scaring the wildlife as well as man, who seeks solitude from the roar of the city. The same damage is done by snowmobiles. Both the Forest Service and Park Service have taken some steps to regulate those vehicles; but the measures are palliatives only.

Come with me and I'll show you a sixteen-mile mountain valley where 10,000 tote goats have been counted on a Fourth of July weekend.

1947227

Come with me to an isolated cabin in deep snow country and hear the snowmobiles roaring. They are now so numerous that it sounds as though a power saw is running full blast during all the daylight hours—right outside your study.

ESTUARIES

Life probably began in the shallow coastal waters of ancient seas two billion or more years ago. There, perhaps with lightning and ultraviolet light as energy sources, simple molecules were bound together into complex structures that eventually became capable of making more of their own kind. In their food habits, these first primitive forms of life resembled animals more closely than plants.

William H. Amos,
The Life of the Pond

Once all life was restricted to the sea. Today even man remains linked to it; our body fluids maintain a constant salinity, and proportions of other ingredients in the blood also resemble those of the ocean. Echoes of the rhythms of the sea are found in the reproductive cycles of many land animals, including humans.

William H. Amos,
The Life of the Seashore

All the life of the shore—the past and the present—by the very fact of its existence there, gives evidence that it has dealt successfully

with the realities of its world—the towering physical realities of the sea itself, and the subtle life relationships that bind each living thing to its own community.

Rachel Carson,
The Edge of the Sea

THE ESTUARINE ZONES OF THE COASTAL STATES are active battlegrounds where private interests are opposed to the public good. The ecologists know that the estuaries are not only a home for waterfowl but also a sanctuary for at least two-thirds of all fish and shellfish during some cycles of their lives.

The estuary—that once-beautiful and unspoiled area where the salt sea meets the fresh water of outflowing rivers and streams—is unique. For estuaries, in a sense, are nurseries of life, incredibly rich in nutrients carried and deposited there by the rivers, which in turn have picked up minerals and natural soil particles from small runoffs far upstream. Estuaries, properly balanced, provide the right mixture of salt and fresh water needed by many species to reproduce and mature.

Historically, the estuaries were not settled early, for they were generally too marshy for building; sometimes, when nearby land areas were stable, or natural harbors were important to commerce, ports began to flourish near their river mouths. But their importance was largely that of gateways to the interior; few appreciated the key ecological role which estuaries play in stabilizing the whole spectrum of life in both rivers and sea. Even those who profited commercially from the products of the estuary failed to appreciate or take steps to protect it.

Sewage, industrial wastes, and oil have destroyed many miles of those critical waters. Pollution, dredging, and filling mark the demise of much of our marine life.

San Francisco Bay lost all its oysters by 1930 and all its clams by 1948. The Bay shrimp has dropped from 6½ million pounds a year to 10,000 pounds.

About two-thirds of the seafood gathered by Americans depends in one way or another on estuaries. Many produce more human food per acre than the best Midwest farm land. They are the home of furbearing animals and nesting and wintering grounds for waterfowl.

Yet they have become dumping grounds for the refuse of cities. They have been drained to rid the area of mosquitoes. Our estuaries were reclaimed by farmers for agriculture. They were considered wastelands suitable for housing projects, factory sites, amusement parks, and other profit-making enterprises.

Development of estuaries followed mass residential promotions. The water is shallow, say six feet, and the loss is twofold: (1) the loss of surface from the fill, and (2) the damage to the bottom due to the dragline. In Tampa Bay the bottom was "still a biological desert" ten years after "development."

Reefs are casualties in the estuaries and coastal waters—casualties of mosquito spray, sewage from trailer parks, and excavations. When the reefs go, the breeding grounds for mackerel, striped bass, drum, sea bass, and bluefish also go.

Many marshes, which are part of the estuaries and washed and renewed by their tides, are private property. Hence dredging is needed if there are to be marinas, and filling is needed if homesites are to be built. Control of the estuaries to prevent dredging and filling may deprive the owner of use to an extent that results in a taking of private property in the constitutional sense.[1] It is the duty of conservation agencies to protect "ecological" values; it is the duty of courts "to stand guard over constitutional rights."[2]

As this is written, battles are being waged in various state legislatures to regulate estuarine zones. Massachusetts and Connecticut led the way in the late sixties. Maryland followed in 1970. Of these, Connecticut is in many respects the pacesetter, regulating private uses under a permit system, and where that regulation results in a "taking" of property, awarding damages.

While some states go forward, Texas moves backward. The dredging of Galveston Bay in Texas was under permits from state agencies which acted to produce a few more Texas millionaires from the oyster shell business rather than to protect ancient oyster beds that were the center of a vast network of rich sea life. The *1971 Joint Report of the Interior Committees on Pipeline Study and Beaches* shows how the Texas state agencies, designed to safeguard the environment, are mere minions of the industrial groups dominant in Texas life.

A regulation by the Parks and Wildlife Commission prohibited dredging except where there was a two-foot overburden of silt. When that occurred, there was no substantial injury to oyster production or to marine life because oysters can't grow if there is that much silt. While the old commission had this policy, the oyster exploiters got a new commission that changed the rule. Then there was a skimming off of the hillocks, the protuberances of shells upon which the oysters grow.

The consequences have been shocking. (1) When the hillocks were leveled and the bed of the bay became a continuous level of silt, the oyster fishery was gone. (2) When Beaumont, Texas, ordered oyster shells to pave its city streets, it discovered that live oysters—not shells—were being dredged; and the odors of the decaying oysters in the streets were awful.

California has also taken dangerously backward steps. It is now pumping waters from the northern part to the arid southern part, sending sweet water over the Tehachapi Mountains so that more gems like Los Angeles can be created. That diversion of sweet waters from the estuaries is a death knell. For marine life needs brackish waters in its nurseries.

New Jersey is another state that is moving backward. The Meadowlands near the mouth of the Hackensack River in northeastern New Jersey receives 42,000 tons of garbage each week and services some fourteen municipalities that now want to "develop" it for industrial, commercial, and residential use—popularly known as "tax ratables." New Jersey created an agency for that purpose. Others promoted a sports and racing complex in the area to accommodate the

New York Giants football team; and New Jersey spawned still another agency to promote it. As this book is written battle lines are being drawn anew for one of the most crucial environmental battles of the century.

Escambia Bay in Florida has been ruined by the dumping of chemical wastes from nylon, acrylic and plastic plants. The marine life has disappeared. Business has collapsed for lack of tourists. Homes and land are losing resale values because of the stench. The manufacturers have preempted the Bay.

The oceans are now the world's sink for disposal of wastes. Every nation has rivers that carry pollution to them. The nations that are technologically advanced—England, France, Germany, Japan, Russia, the United States—are the worst polluters. What they now do will be done by oncoming nations as those nations fulfill their desires to be "industrialized." Great risks are involved. The oceans, which make up about 70 per cent of the earth's surface, generate more oxygen through their plant life than do trees, grasses, and plants on the land. The oceans have been indispensable sources of food. Beyond that is their energizing recreational aspect. All the wholesome aspects are now in jeopardy; and the risk is great that in another decade or so the oceans will be useful to man only as highways. There are already many Dead Seas in the ocean where nothing useful to man can live.

The oceans and the estuaries have suffered greatly at the hands of the U. S. Corps of Engineers, which long gave permits to dump sewage, debris, and other pollutants into the ocean. Millions of gallons of liquids and tons of metal pollutants were dumped into the Gulf of Mexico annually. Lake Michigan alone was used as a dumping ground for about seventy different dredging operations. The Corps long permitted dumping in the Atlantic and in other waterways, which meant that they too received almost daily tons of sludge that was highly polluted by industrial wastes and sewage. Its permits for dumping have not until recently involved inquiries into the poisonous and obnoxious character of the materials being dumped. That helps explain the twenty-mile stretch of sludge in the Atlantic off New York City. Companies have barged wastes down the Mississippi to the Gulf and dumped them. Some substances were in drums; some were released loose if they were water-soluble. Sulphuric acid was taken out beyond the Continental Shelf and dumped. Neutralization before dumping would have been more expensive. Dollars, as usual, carried the day.

The annual tonnage of dredging spoils, sewage, sludge, and other wastes deposited in the ocean was enormous: about 9 million tons in the Pacific, 39 million tons in the Atlantic, about 15 million tons in the Gulf. About 14 million tons of dredge spoils were dumped in the Atlantic alone—highly polluted stuff ranging from raw sewage to toxic metal particles. All of this was done under permits from the Corps, which gave consideration only to problems of navigation. The Corps dredged Chesapeake Bay to keep nagivation channels open. It used the dredging to fill marshes and wetlands. When they ran out, it proposed dumping the excavated material on the upper end of the bay. This brought violent protests from conservationists and conservation agencies. No ecological or environmental aspects of the dumping were a concern of the Corps until 1971, when it turned its great talents in that direction.

Establishing harbor lines is part of the duties of the Corps. There are often large areas of submerged lands shoreward of harbor lines and they are part of the estuaries so vital to marine life. The Corps until 1970 did not concern itself with landfills and construction by private interests on submerged lands and tidelands landward of the harbor lines. The ecological impact was disastrous, and stringent controls over those fills and construction projects became necessary, according to the Dawson Committee of the House of Representatives, if we were to save our estuaries.

A third of San Francisco Bay—or 257 square miles—has

been filled in or diked off and is now occupied by homes, shopping centers and the like. Oyster and clam production are ended; only a minimal amount of shrimp production remains. There are thirty-two garbage-disposal sites around the shores of the bay. Eighty sewage outfalls service the bay. Daily, over sixty tons of oil and grease enter the estuary, the cradle of the sea. The Army Engineers were not responsible for the pollution; but they were responsible for the dredging. The National Sand and Gravel Association has the estuaries marked for billions of tons of sand and gravel for the next thirty years. The Corps issued dredging permits; and ten years of dredging, according to the experts, will make that estuary a biological desert.

Under the hammering of Congressman Reuss and the Dawson Committee of the House, which issued scathing reports in 1970, the Corps is showing some signs of vigilance respecting ecological values in the issuance of permits to dredge and/or fill eleven acres of tidelands for use as a mobile trailer park. Judge John R. Brown said that the government was entitled "to consider ecological factors and, being persuaded by them, to deny that which might have been granted routinely five, ten, or fifteen years ago, before man's explosive increase made all, including Congress, aware of civilization's potential destruction from breathing its own polluted air and drinking its own infected water and the immeasurable loss from a silent-spring-like disturbance of nature's economy."[3]

Under regulations issued April 7, 1971, the Corps is now also responsible for issuing permits that regulate "the discharge of pollutants or other refuse matter" into the navigable waters of the United States—which of course includes the marginal sea. But the Corps in issuing the permits is bound by the findings of the administrator of the Environmental Protection Agency respecting "applicable water standards," the protection afforded fish and wildlife, and other standards that may exist under any pertinent Act of Congress.

Washington, D. C., is still far, far away in the mind of the average person. How does one break through the protective layers with which every bureau shields itself? The critical policies are made as administrative fiats by faceless men hidden somewhere in a federal labyrinth.

States in rebellion against these slow but suffocating controls in far-off Washington, D. C., are beginning to take steps to save all or part of their internal rivers as scenic or natural waterways. Maine, in protest against plans of the National Park Service, led the way with the pristine Allagash, protected against pollution by sanitary zones. Oregon followed suit in 1970, making some of its rivers "scenic" waterways protected by sanitary corridors. Virginia in 1970 established the framework of such a system, the rivers included to be submitted later to the legislature. Kentucky, Tennessee, and Washington are promoting like legislation. With the mounting population, riverfront property is more and more attractive. Builders and promoters plan thick settlements along some waterways that will soon despoil them by runoffs from septic tanks or cesspools. Only sanitary zones along each side of the river—such as Maine and Oregon provide—can save these jewels from becoming open sewers.

Single use should be the standard for coastal areas. If industry is tolerated on an estuary, it should be closely confined. Other parts of the estuaries should be reserved for wildlife; still others, for people.

The solution is not locking up the estuaries. Man has a place there. As John Hay recently wrote:[4]

"Land and intimate human use should not be divided. Seen through local eyes, respectful through long acquaintance with all the details in the landscape, the marshes, the estuaries, the creeks and inlets, the grasses, birds and salt pools are a tremendous resource worth far more than billions of dollars. Compared to such a resource, the value of speculative 'growth,' ruining what it feeds upon, amounts to no more than a gutted clam shell tossed into the sea."

MINING

Strip mining for coal shatters the ecological balance with a finality that defies description. Explosives, bulldozers, and giant shovels assault the hills cataclysmically, like the barrages of Verdun and the fiery mushrooms of Nagasaki and Hiroshima.

In their natural state no region in the world is lovelier than the Kentucky highlands; but as "development" of strip mines proceeds, they become as ghastly as the rotted face of a corpse.

HENRY CAUDILL, "THE MOUNTAINEERS IN THE AFFLUENT SOCIETY,"
National Parks & Conservation Magazine, JULY 1971

Unless men took minerals out of the ground and made them into useful products, we would have no civilization at all. There would be no heat, no light, no policemen on the street, no school teachers.

Mining Congress Journal,
NOVEMBER 11, 1971

AN OLD MINING LAW dating back to 1872, when the open spaces of the West seemed endless and several areas were still Territories, gives almost unlimited entry into public lands (including national forests and grazing areas) to make mining claims. The establishment, maintenance, and validation of these claims under this law is easy. Even as respects "wilderness areas" (set aside under the 1964 Wilderness Act), the same mining rights are preserved until 1983.

One effect is the ancillary right to build roads, to put in heavy machinery, to create open pits for mining and the like. The net result of the mining may be enormously productive and profitable; or it may involve only marginal ore production, which nonetheless can be profitable if the price is right.

Prospectors have another, and sometimes primary, purpose, and that is to sell stock in dubious or marginal projects. The promoter may be working for very little, or it may be one of our giant companies. Whatever the motive the end result is ruination of a wilderness or landscape.

Our mineral resources, which preexisted human life on this planet, are nonrenewable. While population can increase, these mineral resources are not capable of growth or expansion.

By reason of the 1872 Act private persons having claims can work them within the 104 million acres of Forest Service lands. Up to the end of 1971 the Forest Service had no records of what claims were being worked; nor did any other federal agency. Claims must be filed in the county courthouse; but only backpackers and trail riders report the activities of mining claimants.

The advent of new techniques has led to mining on a much vaster scale than ever dreamed of by those who framed the 1872 law. The forty-niner with his pickax and pan has given way to the power shovel and the conveyor belt. The ugly effect of this accelerated assault upon both the natural environment and the nonrenewable resources is nowhere more evident than in two types of mining widely prevalent today: the strip mine and the open-pit.

Strip mining in Appalachia produces 40 per cent of the coal that is mined there, 48 per cent of Kentucky's coal production, and nationwide 44 per cent. The percentages promise to increase in the 1970's and 80's.

Formerly, small power shovels of eight cubic yards were used in strip mining. Today a 220-cubic-yard shovel is more likely; and it can strip off the topsoil (overburden, it is called) to a depth of over 160 feet. Present techniques allow strippers to dig 185 feet beneath the surface; but the technicians anticipate that soon they may possibly be able to dig as deep as 2,000 feet. Today the scoop, known as Big Muskie, is as wide as an eight-lane highway, stands ten stories high, and takes a 325-ton bite. With the arrival of these new shovels, old strip mines are being eyed for renewed stripping so that deeper layers of coal may be mined. That is part of the reason why a drive is on to ban all strip mining.

Strip mining of coal is a dramatic illustration of land abuse. Compounds of sulphur are commonly associated with coal. By surface mining, sulphides are exposed to air and water, oxidizing to produce sulphuric acid, which in turn seeps into streams in runoff and ground water. Even small amounts of acid mine-drainage limit biological productivity and eliminate plant and vegetable life. Vegetation dies off, erosion follows, and the life of the stream is destroyed.

In Maryland and West Virginia the drainage of acid from abandoned mines into the North Branch of the Potomac is so great as to ruin the Corps' proposed Bloomington reservoir for recreation. The acid would corrode aluminum and eat out nails in wooden boats. Swimming or other bodily contact would be perilous. The much-touted recreational potential turns out to be a fraud.

An entire area may be ruined by the strip mine, not just the streams. Since all topsoil and vegetative cover is ripped off, the stripped area has no timber and no plant growth to protect it from erosion by water and wind. In areas with any amount of rainfall, the sediment yield is multiplied enormously, leaving the mined areas bare, causing landslides, and choking the watercourses.

Research conducted in Kentucky indicates that sediment runoff from coal strip-mined lands can be as much as a thousand times that of an undisturbed forest. During a four-year period, the annual average sediment runoff from Kentucky spoil banks was 27,000 tons per square mile, while it was estimated to be only 25 tons per square mile from forested areas.[1]

The alliance behind strip mining is powerful—United Mine Workers, National Coal Association, electric power utilities, and oil, coal, and steel companies. One of their slogans is "Coal means jobs." But as Congressman Ken Hechler of West Virginia says in reply, "Strip mining means temporary jobs." Statistics bear him out, as the fastest exodus of people from Appalachia occurs in strip-mining areas.

The manner of the operation often leaves penniless the owner of the farm whose ancestor sold the "mineral rights" for a song. The Bureau of Mines—a powerful federal agency—has sat on its hands for years while this despoliation went on. The biggest strip-miner today is TVA— the agency that belongs to the public. TVA, the agency many thought was the answer to exploitation, has turned out to be the greatest despoiler of all.

Anger of the people against TVA mounts in Appalachia as the spoil banks increase and lovely mountain country is turned to rubble. Meanwhile TVA power plants, burning stripped coal, cover the area with smoke palls and fly ash.

Some two million acres of land have been devastated in this nation by strip miners whose goal is to get the coal, get the profits, and get out. By 1980 there will be at least three million more acres torn up by the strippers.

The scene is so devastating that Wendell Berry has written[2] that "strip mining is the most enriching to the rich and the most impoverishing to the poor; it has fewer employees and more victims. . . . By the time all reclaimable mine lands are reclaimed and all the social and environmental damages accounted for, strip mining will be found to have been the most extravagantly subsidized adventure ever undertaken." "Jeremiah," he says, "would find this evil of ours bitterly familiar:

> 'I brought you into a fruitful
> land to enjoy its fruit
> and the goodness of it;
> but when you entered upon it
> you defiled it
> and made the home I gave you
> loathsome.' "

The enormity of the exploitation is evident from the fact that although the large coal companies own 33 per cent of Appalachia, they pay less than 4 per cent of the property taxes.

Strip mining is usually identified with the Atlantic states, most notoriously West Virginia and Kentucky. Henry Caudill of Kentucky wrote:[3]

"This forest was here when the Rockies rose up and when they went down and when they rose up again. It has witnessed two great sieges by glaciers but it could not withstand a single assault by the Mellons."

West Virginia, Tennessee, Pennsylvania, have experienced like disasters.

But this desecration also takes place in Kansas, Illinois, Ohio, and now in many regions west of the Mississippi. Thousands of acres of good cropland were made sterile in southeastern Kansas, and bleak ridges of shale and clay, interspersed with pits and trenches, were left behind. Ohio, Illinois, and many states west of the Mississippi with relatively flat areas and great beds of coal seem destined to the same fate. Montana, North Dakota, Wyoming, and six other Western states contain over 53 per cent of the nation's coal reserves;[4] eastern Montana alone has an estimated 3.1 million acres of land containing coal easily accessible to stripping[5] and has been called a "potential strip-miner's paradise."

The Coal Mine Health and Safety Act of 1969 provides for health and safety standards for all coal mines, including strip mines. Included are the noise standards established by federal law, governing factories which manufacture or furnish supplies to any agency of the United States, already mentioned. But the federal government has not yet entered the field of regulating strip mining per se.

In British Columbia a Kaiser Company is strip mining a portion of the Rockies to get coal for export to Japan, removing by huge machinery an overburden of earth to depths of 480 feet. It's called "scalping mountain tops" and leaving "rubble piles" for the future generations. Strip-mining laws in Canada are as weak as they have been in the United States.

Strip mining is done on a huge scale in New Mexico and Arizona. One operation in Arizona is under a ninety-nine-year lease under which the operator hopes to mine 112 million tons in thirty-five years. The coal will be used in California, Nevada, and Arizona to produce power. Much of it will be "slurried" in an eighteen-inch pipeline to Nevada, the water being obtained from deep wells. The Hopis and Navajoes, whose coal is being strip mined, protest the use of the ground water, which is essential in their arid land for irrigation. And the protests extend to the waste water discharges, which are high in sulphur and will pour into the already heavily polluted Colorado. Can it be made mandatory for the water to be evaporated, leaving the sulphur in sealed-bottom ponds? These were some of the controversies raging in 1972 in the Southwest. More will erupt because the land west of the Mississippi now has 77 per cent of all remaining strip-mine sites in the country.

Strip miners in the Eastern small states are penalized for

failing to restore the land. But the fines are nominal; and they often prefer to pay them rather than to undertake the work of restoration.

Some of the Western states have made attempts to carry restoration a step further by imposing a positive duty to restore the land to *productive* use. But these laws are vague, leaving much discretion to the coal mines. In Wyoming, the coal operators are required to seal exposed seams with plant cover only "when practical"; there are no penalties for failure to do so. The Montana statute speaks in terms of returning the land to "useful production," but never defines useful production.

The intervention of government in some more effective form is needed where man has cruelly despoiled the land as he has in strip mining; but legislatures, thus far, have not been successful in dealing with the problem. The laws have been weak and, even so, disregarded by the mining companies. When Governor Combs took office in Kentucky in 1960, he discovered that only a few of the many mine operators in the eastern part of the state had even bothered to obtain a permit as required by the legislature. Kentucky's strip-mining laws in the mid-1960's were so feeble that the Louisville *Courier Journal* won the Pulitzer Prize in 1967 for its criticism of them.

Kentucky has made some attempt to amend its strip-mining laws; but these still fail to recognize fully the difference between "area" and "contour" strip mining. "Area" stripping involves the digging and filling of trenches on relatively flat land; "contour" stripping is done on hilly or mountainous country, and involves tearing down part of a mountain and dumping or pushing it away. Kentucky was able to improve its flatter lands by requiring the mine operators to grade off their rubble piles and replant them; but similar requirements to restore the contours of a mountain by regrading and repiling have proved literally impossible to carry out. Now the drive is on to abolish strip mining entirely in eastern Kentucky, the mountainous part of the state.

Ohio proposes to abolish the practice entirely. In 1971 a campaign was also under way in West Virginia to outlaw strip mining and it was led by John D. Rockefeller IV, who called strip mining "a spreading cancer." Though it is a $200 million a year industry in West Virginia, he proposed to banish it "completely and forever." The legislature went part way in 1971—strip mining was banned in 36 of the 55 counties and new strip mining was limited in the other 19. But the political battle to abolish it completely continues.

There are some coal men, on the union side, who maintain that if all the laws for reclamation of strip-mined lands and for pollution control of those operations were enforced, underground mining would be much cheaper.

There are over twenty strip-mining bills before Congress. The strongest is sponsored by Congressman Ken Heckler of West Virginia. It would phase out all strip mining within six months, provide matching federal funds to states for reclamation of land already spoiled, allow class actions to enforce the law, put its administration in the new Environmental Protection Agency and take it away from Interior, and give states one year to submit environmental control laws for deep underground mines.

Legislation alone, however, is not enough. Fair and firm law enforcement is necessary.

There are a few feeble signs of hope: down in southeast Ohio some spoil banks have been planted with trees or with vetch, ponds have been stocked with fish, and desolate strip-mine areas have been converted into campsites.

Near Girard, Kansas, a new recreational area has been created out of a place made utterly desolate by strip mining.

In a few instances, some of the mining people themselves are beginning to admit that reclamation of mined lands is a serious problem. But they are not prepared for it. They lack people schooled in the life sciences, ecology, agriculture.

But the reclamation projects are minor. Moreover, the

so-called restorations of strip-mining sites are largely public relations facades. The acids that ruined the streams permeate and ruin the new ponds that have been dug. Acid mine-drainage makes a mockery of almost any effort to "reclaim" the ruined earth.

Strip mining in Appalachia has produced many millionaires, but through silt and acid it has ruined 12,000 miles of streams. The ordinary people of Appalachia are sitting on top of a gold mine but starving.

The strip mining of coal that has started west of the Mississippi will dwarf anything the East has seen. It is expected that before long 300 million tons of coal a year (one-half of total U.S. production in 1970) will be strip mined west of the Mississippi. It is expected that this coal will be used in part to make gas that is practically pollution free. Since 1970 speculation in leases for coal exploration on federal lands—national forests, grasslands, deserts, and open range—has been rampant. With the use of new powerful machinery the cost advantage of strip-mined coal over deep-mined coal is three to one; and worker productivity runs five to one. In addition, the process of manufacturing gas requires vast volumes of cheap coal. The West now has most of the strippable coal reserves. Its coal, which is low in sulphur, has increased in value, as urban pollution abatement laws have been enacted that forbid the burning of coal or oil containing more than one per cent sulphur by weight.

The federal agencies leasing these lands are lackadaisical when it comes to setting standards that will protect the earth from being despoiled. The prospect is ominous but real that these new massive strip-mines will create hideous badlands, the like of which man has never seen. The great strip-mining drive in these new dimensions is actually under way and is appalling in its impact. It is almost too late to join all the hands necessary to protect the good earth against the awful oncoming desecration.

There will be thousands of acres of windrows of spoil

banks. Silt will fill streams for thousands of miles. Long-lasting trickles of sulphuric acid will pollute all streams and kill all aquatic life. From the air, pools of rainwater will glow red and orange from the acid they contain. The rubble left behind in this mad search for wealth will have the mark of desolation and despair on it.

The plea that we are short of coal has a hollow ring. For in 1968 we exported 51,220,000 tons of coal and in 1969, 57,034,000 tons.

Drastic preventive measures are needed at once.

All strip mining in an area that produces sulphuric acid to poison the land and the water must be banned.

All mining should be banned unless, as W. T. Pecora, former Director of the Geological Survey, says, "It is rich enough to support proper restoration and re-utilization of the land."

Mining must also be banned (1) where the slopes are so steep and the rainfall so great that reclamation would be impossible or impractical; or (2) where the area has a choice scenic beauty, or where it is important to wildlife; or (3) where the area is so populated that strip mining would be a disruptive force.

Strip-mining areas that are already ruined should be acquired by the federal government and reclaimed.

Strip mining, if allowed, should be sanctioned only where total reclamation can be carried out efficiently and promptly.

Strip mining, if allowed, should follow prescribed procedures; the topsoil should be scraped off and set aside. And, when the mining is over, the topsoil should be replaced in its natural order and compacted, and the land fertilized and planted with trees or other vegetation indigenous to that area.

Strip and surface mining is, of course, a problem not confined to coal. The development of bigger and better mechanical scoop shovels—monsters that can remove tons of raw earth daily—and of huge conveyor belts which now move these materials with ease has led, particularly in the copper industry, to the gouging out of enormous holes known as open-pit mines.

The 1967 plans for the Anaconda Twin Buttes mine in southern Arizona (now operating) called, for example, for an open pit 1,800 feet deep and one mile by one-and-a-half miles at the surface. To get ready for mining, "Anaconda [had to] strip 300 million tons of allurium interlayered with scattered caliche beds to reach the host rock 460 feet below the surface. The stripping . . . require[d] . . . the removal of 270,000 tons of overburden [topsoil] a day."[6]

The Kennecott mine at Bingham Canyon, Utah, which claims to be the world's largest man-made open pit, expanded to engulf the entire town at Bingham Canyon, whose population was formerly 10,000. A few lone protesters objected to the demise of the town on the ground that there should be no form of private corporation, no matter how big, powerful enough to liquidate a government, no matter how small. But the mining company ultimately prevailed.

In addition to the air pollution problems caused by smelting, the tailings pond at the Bingham Canyon mine has caused another kind of pollution. Dust from the vast amount of fine materials in the tailings is picked up by the wind and moved for many miles.

The open-pit mine may sometimes be a whole series of pits, as in the Kennecott complex near Ely, Nevada. Five huge pits yield 22,000 tons of ore and 80,000 tons of waste daily. The ore is sent by rail to a nearby concentrator-smelter plant, where it is crushed, ground and chemically treated to produce a copper concentrate—creating, in the process, noxious fumes which spread for many miles. The new scars upon the landscape are so enormous they can be seen for miles away.

A description of this Nevada mine is in a tourist attraction leaflet entitled "Welcome to Bristlecone Country." The leaflet gives equal space to the mine and to the rare and ancient bristlecone forest in the nearby mountains and is sponsored jointly by the Forest Service, Bureau of Land Management, White Pine County Economic Development Committee, and White Pine County Chamber of Commerce and Mines.

Which interest will prevail: the short-run needs of development and mining expansion promoted by the commercial interest, or the long-range goals of minimizing damage to the ecosphere?

The same leaflet also explains that most lands in White Pine County are owned by the public. But the old 1872 Act gives private mining claims priority over the public interest.

A clash of values and conflict of goals has thus emerged.

A huge ad, sponsored by the Arizona Mining Association, carries the slogan "From copper, a better way of life." The ad lists a few of the many man-made products: air conditioners, sewing machines, stereos, even mine hoist-motors, which are powered by electric machines fused with copper. The defenders say that but for open-pit mining the United States will be out of copper.

Yet that is not true. Most of the copper produced from ore and built into a product could be recovered and used again when the product outlives its usefulness. We do that with gold and platinum. We can do it with copper. Recycling, which I have already discussed, is the ready answer. Depletion of metal is governed not by the amount of metal used, but by the value which we place on it.

The mining lobby is strong; but the conservationists, some federal agencies, and a few states are now lining up in opposition to it. Several states, in an effort to control the noxious fumes of the smelters which are an integral part of some mining-mineral processes, have already set much stronger air-pollution standards than has the federal government. And in Arizona, a recent statewide poll showed that 86 per cent of the people questioned supported the bill of Congressman Morris K. Udall of Arizona, requiring mining companies to get permits before doing mineral exploration on public lands.

Moreover, Congress, in the Environmental Policy Act,[7] states the national policy: the assurance for all Americans of a "safe, healthful, productive and esthetically and culturally pleasing surroundings" and the attainment of "the widest range of beneficial uses of the environment without degradation, risk to health or safety, or other undesirable and unintended consequence"; and it directs that the "public laws of the United States" be "interpreted and administered" in accordance with the Act's policies. As made clear in the legislative history, mining was one target of the new Act.

What effect these directives will have on the ancient mining laws and on entrenched practices such as strip and open-pit mining is yet to be known. But the lines for resounding battles are now drawn.

"When a giant coal company decides it will strip mine because strip mining cuts costs and increases profits, it does it; no Appalachian government, no Congress, has yet been able to determine that preserving land has a higher value than making profits."[8]

WILDLIFE

Vulgarity, in all its varieties, is an attribute of civilized man, an offense not so much against conventional as against natural propriety. It is an expression of revulsion against propriety, against a too-delicate sense of propriety in others and in oneself, against the tyranny of "good taste." It is designed to blunt the edge of such sensitivity. Therefore it is an expression of man's spiritual restlessness, uncertainty, and dissatisfaction. If a redstart revolted against its own instinctive behavior, which represents a binding propriety, it might be vulgar and make a virtue of vulgarity; but it is always itself, not a half-creation like man but a completely realized ideal, a finished product.

LOUIS HALLE,
Spring in Washington

Under the circumstances, one wonders that any shorebirds survived into the present century. Not only were they trapped and shot, but great numbers of knots and other species were taken by "fire-lighting," a nocturnal practice much in favor on Long Island's Great South Bay and elsewhere, in which the resting flocks,

blinded but undismayed by a bright beam, stood by patiently while market men stepped forth from punts and wrung their necks.

PETER MATTHIESSEN,
The Shorebirds of North America

GENESIS 1:28 STATES: "Be fruitful, and multiply, and replenish the earth, and subdue it: and have dominion over the fish of the sea, and over the fowl of the air, and over every living thing that moveth upon the earth."

We make the industrialist the villain. But the naturalist of last century was apt not to be an ecologist. In 1894, the American geologist and anthropologist W. J. McGee extolled the virtues of man: (1) "in the subjugation of the animals of the earth, men preserve only those that can be enslaved, and all others are slain"; and he added that man showed his virtue (2) "by transforming the face of nature, by making all things better than they were before, by aiding the good and destroying the bad among animals and plants and by protecting the aging earth from the ravages of time and failing strength."

We have witnessed a chapter in man's dominion over wildlife that has resulted in dozens of species becoming extinct and several hundred being marked for extinction. The violent attitude toward other life around us has done much damage, upsetting the balance in nature. Ecology teaches that each form of life is part of the same web as all other forms of life. The balance can be dangerously upset when Genesis 1:28 is taken literally.

The Forest Service, the Fish and Wildlife Service, and the Bureau of Land Management are the great destroyers. Congressman John P. Saylor of Pennsylvania called the role of the Bureau of Sport Fisheries and Wildlife in Interior in the killing of wildlife "sinister and contemptible." Another astute observer in referring to the Fish and Wildlife Service said, "Seldom in history has a government agency spent so much time and energy in the official rapine of its own lands, and seldom has the rapine been carried out with such dedication, with such gusto and verve."[1]

Some say that predator control is needed to keep game animals alive for the benefit of man, the hunter. That is nonsense. Paul Errington's studies show that among higher vertebrates predators rarely have a depressive influence on their prey. Predators indeed live on surplus prey, just as a capitalist lives on interest or dividends. Starvation, winter stress, habitat deterioration cause the decline of deer, caribou, or elk—not coyotes, wolves, or bear.

Stockmen have controlled the federal poisoning program. Their associations indeed matched the four million dollars Congress appropriated to the poisoning program. Stockmen almost invariably blame the coyote for their woes. But the coyote is our nontoxic pesticide that eats rabbits, insects, rodents, and carrion. And he is as American "as Indians, cowboys, flapjacks, Bull Durham, sowbelly and beans." Jack rabbits comprise 83 per cent of the coyote's natural food; but man has practically wiped out the jack rabbit. So the coyote is pressed to find food. The coyote is blamed for the lambs that die—even those born dead, those who die because their mothers have insufficient milk, those that are sick or weak. Down in New Zealand, where there are no natural predators, the sheep losses are higher than in the United States. We have eight million sheep on public lands. The grazing rights are for nominal fees. The public lands are fenced and reseeded for the benefit of sheepmen. And the predators are slaughtered in wholesale quantities for the benefit of sheepmen.

There is the "coyote getter"—a device with a pipe protruding above the surface of the ground that is set off by a slight touch and sprays cyanide over the spot—getting little boys, or men, or coyotes or whatever passes. The poisons officially used also include arsenic put out in honey buckets, bait carcasses impregnated with thallium, strychnine concealed in sugar-coated pills, and the nefarious poison 1080.

The whole public domain is filled with these poisons.

They're poisoning wildlife—weasel, mink, fox, badger, coyotes—faster than the animals are born. And carrion-eating birds—eagles, magpies, Canada jays, Clark's nutcrackers, woodpeckers—feed on poisoned bait and go away to die.

Up until February 8, 1972, the only federal restriction on the deadly poisons was that the labels be registered with the Department of Agriculture.

Congress in 1969 enacted the National Environmental Policy Act which requires all federal agencies in their decision making to consider what "impact on man's environment" their decision will have. By Executive Order signed February 8, 1972, the policy of that Act has been applied to the federal poisoners of our public lands. Now they can only trap or shoot predators.

Come with me to Wyoming and I can take you on a horseback journey or on a back-packing hike and we'll not see a living thing in the woods. They are now desolate except for man and his cattle and his sheep that are stomping them to death.

Many states also have predator-control programs sanctioning the killing of bear, mountain lions, mountain beaver, bobcats, and pocket gophers.

In 1970 animals known to have died at the hands of hired hunters were as follows: 89,653 coyotes; 20,780 lynxes and bobcats; 19,052 skunks; 24,273 foxes; 10,078 raccoons; 7,615 opossums; 6,941 badgers; 6,685 porcupines; 2,779 wolves; 1,170 beavers; 842 bears; 294 mountain lions; and 601 other animals. The grand total of larger mammals was 190,763. And these, of course, do not include eagles, hawks, owls, prairie dogs, or black-footed ferrets.

In 1971 Congress finally prohibited shooting any "bird, fish, or other animal" from an aircraft or using an aircraft to harass any "bird, fish, or other animal."[2]

The lobbyists of sheepmen demand ever-increasing killing. The only "reform" is a pledge by sheepmen not to use the lethal agents on their own but to have them used by the Bureau of Sport Fisheries and Wildlife. As Michael Frome has written,[3] "No Federal agency can possibly be more abject in its 'clientism' to special economic interests than the

Bureau in dealing with the sheep industry."

In Teton County, Wyoming, the tax levy for predator control is four times as high as the levy for the county library.

In Wyoming there were thirty-four stations containing the poison 1080, all within a fifty-mile radius of Yellowstone National Park. Wildlife roams widely, not aware of boundary lines. Those who ate off the carcass of an animal killed by 1080 also died. The impact on eagles was severe.

EPA followed up the President's Executive Order banning the use of poison on public lands by banning the shipment of these poisons in interstate commerce. It also announced a complete halt to the use of thallium sulfate, the poison that probably killed twenty eagles in Wyoming in 1971. And the ban extended to cyanide, strychnine, and the notorious and deadly 1080. But the drive against predators continues. Shooting and trapping of them goes on unabated. The stockmen claim that shooting and trapping are more expensive than poisoning even though Washington, D.C., has assigned hundreds of federal agents to help out. Moreover, some states are allowing the use of poisons on private lands.

Interior permits golden eagles to be killed in thirty-three counties in Texas. Interior is shooting grizzly bears. Interior allows female seals to be taken in the Pribilof Islands in the north Pacific even though for every female taken three die: the mother, the embryo, and the pup left behind on the beach.

Shooting eagles from aircraft was long a pastime in Texas. Though it is now banned, some ranchers occasionally hire pilots to do the dastardly act. Nearly eight hundred eagles were shot in Colorado and Wyoming by one pilot in 1971; and ranchers have been baiting, trapping and shooting eagles in thirty-two Texas counties under Secretary Udall and in thirty-three Texas counties under Secretaries Hickel and Morton.

One helicopter pilot operating over Wyoming saw over five hundred eagles—both bald and golden—shot from the aircraft. Most of the birds were flying. A few were shot from the aircraft while they were feeding on the ground. Ranchers paid $25 for every eagle shot and $50 for every four-legged predator or sometimes $80 an hour. Shooting eagles from an aircraft is a federal offense. So is poisoning them. Yet the shooting and poisoning both in Texas and in the Rocky Mountain area have gone on apace.

The culprits are on government subsidies for growing or not growing animals. They are at least in part on public lands under permit from the federal government. Why not revoke the subsidies or the permits of all law violators, including those whom Jack Olsen calls members of "the poisoning establishment"?[4]

Many think that the federal government subsidizes pollution of air and water by the contractors it hires. Some 700,000 procurement contracts are made annually and a sizable percentage end up violating state or federal water quality standards. Federal contractors under existing laws must meet certain wage requirements and conditions of work. Why not require them to comply with local and federal requirements for the protection of water, air, fisheries, and wildlife?

The Department of Agriculture encourages the drainage of ponds where ducks breed so that they may be reclaimed as farm land. But the ultimate aim is not food production, but the payment of benefits to the new farmer who, after a few years of operation, takes the land out of production.

In 1970 the Water Bank Act provided a partial antidote to that practice. Now the Secretary of Agriculture can enter into ten-year contracts with owners of fresh-water wetlands: the owners agree to maintain them as such, and the Secretary in return makes annual subsidy payments to them.

Out in east Texas only about a dozen ivory-billed woodpeckers remain; and now there are many hunters anxious to get them since they are rare, practically extinct specimens.

Technology exacts a heavy price from wildlife. Unethical

pilots hunt golden eagles and wolves from the air. Flesh and blood cannot withstand that assault. Even when the snow is only a few feet deep, deer and elk can be endangered by machines. In the winter when deer and elk live a precarious life, snowmobiles reduce their prospects of survival. Merely chasing them may be fatal at a time when life is at a low ebb.

In parts of New England, beaver are being seriously depleted by trappers who race from trap to trap with snowmobiles. Hunters track bobcat by snowmobiles and then turn loose the dogs, who have been towed on sleds. Fox are chased to exhaustion by these machines.

There is much to be said for the predators. No predator but man has ever caused the extinction of a species.

Stockmen like to turn their sheep loose in the plains country without herders or without constant herding. Losses to predators, if any, are probably more than offset by savings on the herders.

Certainly the presence of predators is a reliable gauge of the balance of nature and the wise use of our grass- and timberlands.

Wildlife gives man values which sheep do not. It is desirable to end all the killing of predators and to install instead a system whereby a sheepman is compensated by the government for every animal which he proves was killed by predators.

FOREST AND WILDERNESS

The human body is a magical vessel, but its life is linked with an element it cannot produce. Only the green plant knows the secret of transforming the light that comes to us across the far reaches of space.

LOREN EISELEY,
The Unexpected Universe

No doubt Saint Paul was right to preach against the people of Antioch, and other prophets to lay their curse upon other cities. But they did the right thing for the wrong reasons. Those sins were not moral; they were not theological—they were ecological. That pride and that luxury might have been a great deal more pronounced and yet no harm befallen them; the green fields would have continued to yield them increase and the limpid water to bring refreshment; that immorality and that impiety might have spread further and mounted higher, and still the strong towers would not have shaken and the massive walls would not have crumbled; but because they had been unfaithful to the land upon which they lived and which God had given them; because they had sinned against the laws of earth, and despoiled the forests, and loosed the floods, they were not forgiven, and all their works were swallowed in the sand.

JOHN STEWART COLLIS,
The Triumph of the Tree

It is my contention that blue plastic toilets, especially when placed in the center of a natural prairie, do essentially alter the primitive conditions and most certainly do use up a hell of a lot of taxpayer dollars.

L. J. COSTELLOE,
"THE BUREAU BUNGLED,"
Oregon Outdoors,
NOVEMBER 1971

THE FEDERAL GOVERNMENT OWNS ABOUT 34 per cent of our total land acreage, including parks, grazing lands, semi-arid lands, and national forests. All except the parks are commonly leased to private enterprise. Highway, airport, reservoir and flood-control projects consume about 580,000 rural acres a year. Uncle Sam is slowly chewing up our out-of-doors and putting it either under water or under concrete.

The Park Service so "civilizes" national parks as sometimes to create smog. That is one threat to Yellowstone National Park today.

The Park Service, by the use of funiculars or roads, likes to pour masses of people into the remotest wilderness rather than save those sanctuaries for the adventuresome.

The Pickett Range—the wildest of our mountain areas short of Alaska—was included in the North Cascades Park. Only 2,000 people a year visit it, as one must go on foot or horseback. The Park Service sold its claim to the Pickett Range when it boasted it could put 20,000 or 200,000 a year into those mountains—as it can if roads or funiculars are built.

The truth of the matter is that the driving force behind the Park Service is the concessionaires. They have a financial stake in "mass movements" into the parks and encourage all devices that promote that end. They are indeed the favored few, for they are the only ones who can receive under Park Service rules notice and hearings of changes in accustomed ways of doing business. They are favored because "mass recreation" is favored. The Park Service loves its "mass recreation" statistics, and promotions are made on that basis. The hikers, campers, and environmental groups are bothersome outsiders who have no right to notice and hearing. The profit motive conditions all Park Service activities.

National parks are "overvisited, overcrowded, overlittered, and overpolluted," as Michael Frome said in *Field & Stream,* March 1972. The drive for urbanization of the wilderness is a curse of national park management.

The national forests are largely considered to be croplands. Virgin timber, sorely needed for wilderness areas and for watershed protection, is being almost daily cut with Forest Service permission, much of the lumber being shipped to Japan.

Buffer areas around true wilderness areas are being sheared by the Forest Service, which means that roads will now reach the edges of the "wilderness," which means in turn that jeeps and tote goats will roar into the sanctuary and tear it up.

Exhibit A is the Goat Rocks Wilderness Area in the Cascades, which lies at or above the tree line. Jeeps and tote goats can reach it easily, and there is no Chinese Wall to stop them once they reach the border. They roar into the wilderness area; and their guns have killed every living thing in that wilderness—every squirrel, every marmot, every camp robber. There are not enough federal personnel to stop them.

The Forest Service is now cutting timber up to 10,000 feet, where it takes 400 years and more to grow a tree. Even at 2,000 feet it takes at least 180 years to grow a merchantable ponderosa pine in the Pacific West.

A 1970 report of the Montana School of Forestry points out that in the Bitterroots the Forest Service cutting is on steep and rugged terrain where it is uneconomical to grow new stands. The Forest Service, it says, is engaged not in "timber management" in the Bitterroots but in "timber mining."

Clear-cutting the forest—in which every tree in huge areas as large as a section of land is cut—was the practice in most areas and, where pine was involved, the clear-cutting was followed by burning the slash to promote regeneration; but that exposed the mineral soil and resulted in great erosion. Where clear-cutting was in Douglas fir and hemlock, the new undergrowth included many berries that bears found attractive. Yet bears also occasionally strip the bark off a young sapling for its sweet cambium layer. Hence lumber companies hired hunters, who killed most of the bears. The Montana Report said that "Consideration of recreation, watershed, wildlife and grazing appear as afterthoughts" to Forest Service management.

We can use the lower reaches to grow trees as we grow asparagus. But the heights must be reserved as sanctuaries.

The agency, entrusted with protecting the public interest in national forest lands, has become lumber-minded and has lost sight of "the entire spectrum of forest-related values," to use the words of Senator Gale W. McGee of Wyoming. Big and small lumber companies hold great sway over the Forest Service. The training of foresters orients them to lumber and logging. The pressure group they see every day is the timbermen. Their local (district) offices are close to the community and sensitive to the desire of even the "gyppo," or small, speculative logger who, if he can cut Sunrise Creek, will put eight men to work. Lumber companies often reward "cooperative" Forest Service personnel with the lucrative executive positions on retirement or even sooner. The Forest Service is so wedded to the lumber industry that public protests, if they are to be successful, must produce a veritable gale that blows through the inbred, commercial-minded Forest Service.

Public criticism of the Forest Service over its lumbering activities has been mounting. The lumber industry now speaks of its "rightful" interest in national forest trees. But they have no "right" to that subsidy any more than sheepmen and cattlemen have the "right" to graze their animals there or prospectors have the "right" to minerals on public lands. The "right" is that of "we the people" who own these forests. The Forest Service for years has been under a mandate to guard the national forests "for outdoor recreation, range, timber, watershed, and wildlife and fish purposes."[1] But the replies of the Forest Service in 1971 to criticism are a self-indictment of mismanagement and a gross abuse of trust. *Forest Management in Wyoming* (1971)[2]—a Forest Service report—says in effect that the uses other than timber production have not been protected. The reasons given are interesting—the lack of trained men, the lack of adequate budget, the lack of understanding of the needs of wildlife, the protection of soils, the improvement of air and water quality, and the promoting of all other multiple uses. Reading this Forest Service apologia will make even a Forest Service fan come full circle and realize that the regimes have been mostly fraudulent and have abused the public trust.

National forests have been considered crop lands; and that is why they are in Agriculture. National forests need to be lumped with national parks and all other public lands so that they can be managed by ecologists. The manner in which Forest Service practices, including clear-cutting, have desecrated the land is shown by the photos included in *Forest Management in Wyoming.*

The forests are very seldom flat bottomlands. In the Far West, they are on steep mountainsides that are ecologically fragile even when covered with stable forests. With cutting we face erosion that is devastating.

There is no timber shortage. Several billion board feet of our timber are exported yearly. Some talk about the needs of the ghettos, as if it is necessary to ruin our wooded lands to save the poor. Housing goals can be met without any increase in timber production. The truth is that lumber consumption is declining. It has indeed been static since 1910 despite our mounting population.

The lumber industry has a great fear of the wood substitutes that are gaining in popular appeal. To enable the lumber industry to hang on to the market we place our forest lands on the sacrificial block. We now think of forestry as being either profitable and therefore desirable, or unprofitable and therefore undesirable.

Congress in 1970 rejected the National Timber Supply Bill.

This bill would have greatly accelerated the cutting of timber and building of roads in our national forests; and would have destroyed many potential wilderness areas. But a few months later, the Executive accomplished the ends of the rejected bill by an Executive Order.

Congress has long declared multiple use as a standard applicable to all national forests. They shall be administered as directed, "for outdoor recreation, range, timber, watershed, and wildlife and fish purposes."[3] And it admonished that "harmonious and coordinated management of the various resources" be achieved in "not necessarily the combination of uses that will give the greatest dollar return or the greatest unit output."[4]

Multiple use has become a misnomer. Land used for a highway preempts all other use. Land used for strip mining is the same. Certain types of logging or even grazing may ruin most, if not all, other uses. The idea of multiple use is to honor all uses when the land is put to a particular use. Watershed values, stream protection, the safeguarding of wildlife, recreational uses by man—these should not be materially impaired by any other dominant use. Yet logging in the Far West has increased the sedimentation of streams, wiping out salmon runs. That is not multiple use. Open-pit mining not only destroys esthetic qualities but pollutes the streams and causes untold erosion by runoffs of water. That is not multiple use. Multiple use has long been written into our laws; yet it has been so perverted by the Forest Service that a new standard is needed.

The Public Land Law Review Commission in its 1970 Report threw its weight on the side of dollars and against ecology, saying that "commercial timber production" should be "the dominant use." That dollar policy has all but ruined our public forests.

We need an ecological standard—stabilization of water supplies, production of atmospheric oxygen, protection of flora, fauna, and topsoils, preservation of outdoor recreation, and salvaging the spiritual values inherent in scenic beauty and the wonders of the woods.

Ecological standards include a selective use of fire. It was fire that kept the hardwoods of the Everglades down, leaving breathing and living space for the Caribbean pine. The Park Service soon realized that without controlled burning the pine would be choked out; and so the Rangers set the saw grass on fire when the wind was right.

It was discovered that the natural conditions for Kirtland's warbler were produced only by forest fire. The bird nests only in large, homogeneous blocks of jack pine varying from five to fifteen feet in height, the crucial requirement being live pine-branch thickets near the ground. This habitat, which is unfavorable to nearly all other forms of life, and which develops after a forest fire, frees the Kirtland warbler from competition and insures its survival. The Forest Service and state agencies in Michigan have combined to preserve this type of habitat in the northeastern part of the Lower Peninsula of Michigan. This warbler never comes to an area opened by lumbering. Mining and other human intrusions are not congenial to it.

Humans are not congenial to the condor. The condor is now reduced to a flock of sixty birds, and they are largely confined to a sanctuary in Los Padres National Forest in California. Next to the sanctuary is private land which developers are planning to convert into small homes and a place for a motorcycle club to operate. Condors cannot stand human intrusion, let alone tote goats. So they seemed doomed. Some animals—like people—need special sanctuaries.

We must practice ecological forestry that respects the ecosystem, not dollar forestry that lines a few pockets with money.

Farm-lot management of forests is a risky thing. The mixed hardwood or hardwood-pine forest is a complex, diverse, and relatively stable association of plants, with a tendency to maintain its ecological norm. As Michael Frome has said:

"True multiple use precludes using forests for growing stands of a single species, as is the case in clear-cutting. Entomologists warn that a pure stand forms an ideal situation for damage from insects and disease; infection is rapid and direct from tree to tree, and, if one species is destroyed, there is nothing left. A monoculturally managed forest creates the need for pesticides and herbicides and for fertilizers that ultimately take more out of the soil than they put into it. The biotic diversity is destroyed."[5]

We are now in an era of scarcity of some species, for example, the black walnut. This tree is prized for veneer and has become very scarce. Some groves still flourish in the Midwest; and they are raided day and night by men who bulldoze a road through another's land, cut down the walnut, and truck it to market.

Clear-cutting of timber is having severe repercussions in Alaska. The choice timber lies along rivers, and clear-cutting of river areas dooms both the salmon and the crab. Erosion pollutes the water; cutting raises river water temperature, and it is known that a 2-degree rise in Fahrenheit means death to certain small fish. This all might not be so bad, Alaskans say, if clear-cutting were essential to their "growth." But in Alaska timber that is clear-cut goes to Japan for processing. It is a form of foreign aid. But Japan is not underdeveloped. It probably will soon surpass us in GNP. Do we need to ruin our salmon streams to make rich Japan richer?

In 1972 the Council on Environmental Quality, headed by Russell Train, proposed an executive order drastically regulating clear-cutting. But the Department of Agriculture, which represents the Forest Service, summoned the timber interests to Washington, D.C., and defeated the proposal.

The case against clear-cutting was never put better than by a logger at a Senate Subcommittee hearing in Portland, Oregon, August 1971:

"My name is Bob Ziak. I am a clear cut, high lead logger.

"I was born in Astoria, Oregon, 54 years ago. My father was a logger. My mother took me from the hospital to a logging camp to live. The forests are my life.

"At first the timber was virgin, production was tremendous, there were no controls and the resulting destruction and waste was appalling.

"I've clear-cut to the edge of a river, destroyed priceless streams, found jewel-like lakes within our cutting lines and left them as ugly holes staring into the skies.

"I've seen the eagle tree left standing all alone only to see the birds leave and the tree die because each needed a stand in order to survive.

"I helped log thousands of clear-cuts, saw the animals move in and then come under a murderous cross-fire from hunters on the network of roads, with no place in sight to go in their terror-stricken flight.

"I am deeply concerned about our forests. They are disappearing—from 600 years of age to 35. Man planted in solid blocks, tree farming if you wish, but our forests are going, going, gone.

"Detach yourselves from this earth and look down on us from the heavens above. Gentlemen, this is all there is. There is no more and the time is running out."

Sea otters, which need clear-running streams, were transferred a thousand air miles from the Aleutians to Chichagof-Yakobi Island wilderness. But the Forest Service in disregard of the needs of the sea otter has decided to log 8,000 acres in the Island wilderness. Logging traffic on the land and in the water promises to seal the fate of the sea otter there too.

We may rant and complain about reckless logging practices prior to the advent of the Forest Service. Yet what was done then was minuscule compared with what goes on today.

Last century lumber companies were cutting 1,300 acres of redwoods a year. In the 1960's they were cutting 13,000 acres of redwoods a year, which averages out at 1,000 redwood trees a day. Once the Congress passed the Redwood National Park Act in 1968, the redwood industry began to cut feverishly—twelve hours a day, six days a week with the most destructive techniques ever known to man. The cutting goes to the edges of the park, so that there will be no incentive to expand the park. The cutting will shortly ruin parts of the park, for the cut-over areas are now denuded, the soil is eroding, creeks in the park are filled with mud, alluvial fans of mud are being created in the park which bury the roots of the redwoods and mark them for certain destruction.

We must put an end to large-block clear-cutting.

We must use selective cutting, small-patch cutting (never over forty acres), and comparable methods aimed at long-term productivity of our forests.

All logging above the 3,000-foot level should be banned. The watershed in the higher country is too important and the land too delicate for it to be disturbed.

The decision to cut or not to cut must by law be made only after public hearings aimed at the question of whether in the given situation ecological considerations and the profit motive are compatible.

We must put an end to the use of herbicides.

Easements must be obtained over large private tracts of timber so that ecological controls are put over them as well as over public lands.

There are many facets to this problem of land use. As already noted, if we recycle only 50 per cent of the paper and paper products in our garbage, we can release 90 million acres of timberlands for recreational, watershed, and wildlife use.

A "wilderness" area as defined in the 1964 Act[6] is a roadless area of at least 5,000 acres. Imagine what could happen if cut-over land were set aside as a "wilderness" area. The mutilated hardwood forests in the unique Big Thicket area of east Texas would in time be restored by nature to their original beauty. The badly bruised spruce-hemlock-Douglas fir country of the Pacific Northwest would in time also be restored. Maine has been cruelly mutilated. So have other portions of the Appalachia, and Ohio, and many points west and south. Temperature, sun, and water can rejuvenate wastelands and provide the basic healing that is needed, provided man stands back and lets Nature take over. Eastern lands will revert to a primitive state within a hundred years; and, though dreadfully managed in the past, they will "quiet down in 20 years or so."[7] This is what George B. Hartzog of the National Park Service has called "recycling of lands," a splendid idea.

The possibilities of reclaiming wastelands or renovating them for new uses are infinite. The need to do so will increase as the population mounts and even remote trails become so crowded with hikers that at long last every campsite in the high country is rationed under a permit system. Indeed the Forest Service in 1971 required permits for one to enter overcrowded wilderness areas in California; and in 1972 the Park Service started rationing the number of people who could visit the back-country areas.

The trails of some wilderness areas should be open to travel either by horseback or by backpacking. Horses, however, present special problems. They may trample a small meadow to death. Some areas have insufficient horse feed, making it necessary to pack horse feed in. That is true, for example, of the Bob Marshall Wilderness Area in Montana. We must soon classify wilderness areas for the type of use permitted.

That is one zoning problem; there are others. Some wilderness areas can tolerate grazing by sheep. Yet that requires close supervision and drastic limitations on the number of animals allowed and the duration of their stay.

Most wilderness areas need to be protected from sheep and

cattle. That is true of the Cascades in Washington and Oregon and the Wallowas in eastern Oregon—all of which still bear the awful scars of unrelenting grazing that reached its peak about 1910. These high alpine basins are fragile; it took thousands of years to produce a quarter inch of topsoil. The sheep trampled the grasses and forbs away and the winds did the rest. Our subsidy of sheepmen—and cattle too—has turned our high basins into deserts.

The drive is on to turn our more fragile desert areas of the Southwest into rubble. Tote goats should be banned from most trails, as they raise havoc with people, horses, and wildlife and chew the trails to dust. When allowed into high basins they churn them into desert bowls in only a few hours. Racing like mad, they destroy the delicate, exquisite beauties of the high country in only a few hours. If allowed, they must be restricted to lowland trails and roads where the damage is already done.

The complete abolition of the snowmobile from wilderness areas or roads leading into them is mandatory. They are impossible to patrol under severe winter conditions and do untold damage.

A few lakes—but only a few—might allow floatplanes. But generally they should be banned from all waters in wilderness areas; and the same ban should be put on helicopters operating in the high country. Government agencies can come in by air on emergency missions. But the early experience with the Quetico-Superior region in northern Minnesota proved conclusively that if private or commercial planes and helicopters are not banned, the solitude of the wilderness will be destroyed and pot-bellied urbanites will fly into lakes in such quantities as to destroy quiet, solitude, and fishing for everyone else.

Mining must be banned from wilderness areas. Mining means roads and roads mean the invasion of vehicles which marks the end of wilderness. Strip mining—which Kennecott Copper plans to undertake in the Glacier Peak area of the North Cascades—is sacrilegious. It utterly despoils a cathedral-like sanctuary.

A coal company has been granted permission by the Forest Service to drill 130 core holes in the Bankhead National Forest near Birmingham, Alabama, to see if strip mining is economically feasible there; and it is considering the issuance of oil and gas prospecting permits in the same locale. Today the miners appear to be the greatest opponents of wilderness areas where there is a prospect of ore or gas or oil. Cabeza Prieta Game Range in Arizona embraces 800,000 acres between Yuma and Ajo. It was set aside in 1939 for bighorn sheep and desert pronghorn antelope and in 1971–72 was being considered for "wilderness" classification. Speaker after speaker appeared at the hearings to describe the "wretched tomorrow" that would arrive if Cabeza Prieta were made exempt from geologic exploration and mining. That was the powerful voice of the mining lobby.

Mining presents a very special problem. The mining lobby is one of the strongest in Washington, D. C. It got into the 1964 federal Act a provision allowing all mining claims within wilderness areas to be staked, proved, and mined, up to 1983. That provision must come out.

Some people in government, and some out, want our mountain retreats developed in the manner of Switzerland. That would destroy the sanctuaries for which this continent has been famous.

But that is the direction in which Big Sky in Montana is moving. It promises to become a cancerous growth in wondrous and pristine country—a place of condominiums and luxury living for the rich. It helps fulfill the prophecy of men like Eugene Rabinowitch, who predicts that man's "ideal is bound to be a growing park rather than a spreading wilderness."

The trend should be toward auto campsites in the valleys and trails and three-walled huts in the back country. That direction—not the one taken by Big Sky—would more nearly assure that the public lands are for those from the ghettos as well as for the rich.

All roads must be kept out of the high country. It must be reserved as the special reward to our great-great-grandchildren who will see it as Daniel Boone, Henry Thoreau, and John Muir once saw it. They should be able to discover and appreciate the America that once was but that the machine has largely destroyed.

Some say that the maintenance of wilderness areas favors the rich and discriminates against the poor, that only by opening up the sanctuaries with roads can the poor receive the benefits of the wilderness that the rich enjoy. But this rich-poor syndrome is a myth. The costs associated with wilderness recreation (apart from skiing) are comparable to or lower than those associated with other outdoor recreation. The costs of wilderness travel on foot are certainly no more than $5 a day. The choice of wilderness recreation turns not on income but on taste preferences. Money does not form tastes; nor does wilderness price even the poor out of the market.[8]

National parks and national forests should have roads only on their perimeters, leaving the inner sanctuaries untouched. All civilization can be contained there. The motels, hotels, and campgrounds on the perimeter can be the takeoff points for those sturdy enough and brave enough to venture into the interior.

No zoning should be done by administrative fiat. There should be public hearings where all those *for* and *against* can be heard. The people own the public lands, and if they choose to turn each park or forest into a New York City Central Park, that is their prerogative. If they so decide, the future will indeed be gruesome. But I have a deep faith that our people will not want to destroy the great American outdoors —an important segment of our spiritual heritage and the greatest bit of outdoors in the whole wide world.

TRANSPORTATION

A city that outdistances Man's walking powers is a trap for Man.
ARNOLD TOYNBEE,
Has Man's Metropolitan Environment Any Precedents?

I should like to have called the attention of the senators on Capitol Hill to these geese overhead . . . In a way, the geese above and the lawmakers below, are earthly travelers that pass each other as if they belonged to other dimensions. There is no exchange of signals, no correspondence, no recognition between them—the aboriginal and eternal wilderness on the one hand, on the other the passing carnival of civilization. How many city dwellers, in their somnambulistic preoccupation, ever know that wild geese are overhead, or violets underfoot in the crack of the pavement? The city is threatened with invasion from every side.
LOUIS HALLE,
Spring in Washington

SOME NATIONS TRAVEL ON FOOT; and some are dependent on bicycles. But we and Europe and Japan are dependent on motors, the primary motor being the internal combustion engine.

Our obsession with motor transportation has given rise to the powerful highway lobby, which is virtual dictator of the freeway system that becomes more extensive and pervasive every year. The agencies that locate, design, and supervise the construction of freeways are in the Department of Transportation. Up until 1970 there were no ecological standards imposed upon the department. Highways were designed in such a way as to ruin trout and salmon streams, at least fifty having been destroyed in recent years in the Pacific Northwest. They were ruined because the gravel in the stream was used in construction, the result being that the old stream became a sterile sluiceway. Or rubble from the excavations was dumped into the creeks, making the water so fast that neither trout nor salmon could negotiate it.

Under Secretary of Transportation Alan S. Boyd there were rules and regulations which put down for public hearing both the question of a freeway's location and the question of its design. The highway lobby was incensed at that interference with its control, and Secretary Volpe revoked the new regulatory scheme. But participation in determining where a freeway should be located and what design it should have is essential if the mad scramble for more and more freeways is to be curbed.

The highway lobby (which now says it does not exist) has loosed a barrage of books, leaflets, and speakers to show what a boon it is to the nation. One sample is that it would take 25,000 years to pave over the entire nation at the current road building rate!!

In New Hampshire, voters at any town meeting can designate roads as "scenic," in which event improvement work may not involve the removal of trees nor the destruction of a stone wall unless, after a public hearing, the planning board approves.

The billboard lobby—like the highway lobby, the mining lobby, the automobile lobby, the industrial lobby, the timber lobby, the livestock lobby—is a powerful force in Washington, D.C. Yet there is a little progress in removing billboards. The "Lady Bird Bill" enacted by Congress ordered states to remove billboards on federally aided highways or face the loss of 10 per cent of the federal funds. But in the six years following passage of the Act, billboards increased 20 per cent and the federal agency did nothing until 1971, when eleven states were notified that they would lose federal funds. Only twenty states plus the District of Columbia and Puerto Rico have complied. When Texas failed to pass a highway billboard limitation Act, $24 million of federal aid to highway construction in Texas was cut.

The law is full of loopholes leaving billboards untouched on about three million miles of secondary roads. While the federal law reached billboards that are 660 feet from the highway, thousands have now been moved back just beyond that zone.

Ogden Nash wrote:

> I think that I shall never see
> A billboard lovely as a tree
> Perhaps, unless the billboards fall,
> I'll never see a tree at all.

In 1956 Congress created the Highway Trust Fund[1] into which all federal excise taxes on motor fuel and automotive products are placed. All of those funds are payable only for building highways that receive federal aid, not for the construction of other transportation media such as subways, monorails, streetcar systems and the like. The balance in the trust fund stays at about 4 billion dollars. Thus for the fiscal year ending June 30, 1971, the trust fund spent nearly 5 billion on highways, yet it had left in it at the end of that time 3 billion 651 million dollars.

With this trust fund the highway builders can go madly on their way—building where they choose and as they choose, holding no hearings, and appearing before no committees of Congress for questioning.

The highway lobby estimates that, between 1970 and 1990, 600 billion dollars, at 1969 prices for construction, will be needed for highways. The Highway Trust Fund is its first article of faith. It helps make the highway lobby indeed supreme. One basic reform therefore is to make the Highway Trust Fund available for all transportation projects, including the development of mass transit systems for those who cannot afford cars. In March 1972 Secretary John Volpe recommended that course, so that beginning July 1, 1973, up to 21 per cent of Trust Fund money could be used for mass transit, and two years later up to 39 per cent.

The federal Highway Trust Fund receives about $5 billion annually. Other billions flow into state highway funds which, like the federal funds, are confined by law to highway purposes only. The poor can go hungry, the cities can decay, the schools may be on reduced budgets, automobile traffic gets worse and worse, the need for mass transit mounts, but not a dime can be diverted from the highway trust.

The efforts to get mass transit have not all been in vain; but they have been opposed by powerful interests. The Nader report *Politics of Law* [2] relates how major oil companies contributed large sums to a committee known as Californians Against Street and Road Tax Trap to defeat the use of state gasoline taxes for mass transit and smog research.

Highways and roads are a curse to a wilderness. In the early days of the nation, lumbering took place without leaving many permanent scars. The roads used to get the timber out grew over and were swallowed up. Today a lumbering road is quickly jammed with jeeps and tote goats; and the mass invasion of people starts as soon as the loggers leave.

That means that wilderness areas must be designed to include wide buffer areas where no logging and no roads are allowed.

Highway planners seem to aim invariably at parks, partly because parks are already owned by the state or federal government, entailing no cost of right of way. But Congress has spoken, saying it wants transportation plans that "maintain or enhance the natural beauty of the lands traversed."[3]

Congress indeed directed the Secretary of Transportation not to permit a federally financed public road to run through a park "unless (1) there is no feasible and prudent alternative to the use of such land, and (2) such program includes all possible planning to minimize harm to such park."[4]

A later Act, the National Environmental Policy Act of 1969,[5] provides that a federal agency, including the Department of Transportation, shall include in every recommendation for any major federal action affecting the quality of the environment, a detailed statement showing what the impact will be and alternatives to the proposed action.

Everyone knows how awful it is when urban sanctuaries —not to mention wilderness areas—are filled with structures, paved with concrete or asphalt, and converted into thoroughfares of high-speed traffic.

Highways and roads are about to ruin our national parks. There are traffic jams in Yosemite and in Yellowstone, to name only two. The highway lobby points out that many of the roads inside national parks do not meet minimum safety specifications. They also point out that existing two-lane roads are simply inadequate to handle existing traffic. The highway lobby therefore wants to build additional highways inside the parks to handle the traffic. The highway lobby in 1971 organized "truth squads" to go about the country promoting the building of highways and attacking the environmentalists.

Our mounting population and mounting traffic mean we must make a complete reversal of our outdoor policies. All cross-park and cross-forest traffic must be discontinued. All cross-park and cross-forest roads must be closed. Roads at their present rate of growth will destroy the parks and the forests. Roads should lead only to campgrounds or parking lots on the perimeter. From these points visitors can explore the interior—on foot, on bicycles, on horseback, or even wheelchair. If there must be any mass movement of people within park or forest boundaries it should be by buses on regular schedules.

As the director of the Park Service recognizes, recreational vehicles commonly known as "campers" have added greatly to the problems of crowding and congestion, and may also have to be banned. Each year the new models are bigger and more elaborate, taking more space, requiring wider and sturdier roads, demanding hookups for water and electricity; with the net result that in some parks the owners of campers literally "park" them, set up their folding chairs and transistor radios, and seldom venture more than a hundred yards away. This kind of "recreation" should certainly be confined to the fringes of the park, leaving the interior to those who simply wish to hike and fish and enjoy the solitude.

Parks and forests cannot bear up any longer under the automotive onslaught. As we have said, autos should take tourists only to the edges of the wilds. It is there that civilization should be contained, saving the interior for those who can find an alcove of solitude in a great sanctuary of nature.

Some public lands need special protection against motor vehicles. It is common in the Southwest for people to take off cross-country; and they are doing irreparable damage. Desert flora exist on a precarious edge. Cacti and other low shrubs are shallow-rooted for quick absorption of any rain. Those plants include the saguaros, chollas, prickly pear, and creosote bushes. They are easily uprooted. Moreover, when rain comes, it is often torrential. A tire rut can quickly become a gully and a place of beauty is destroyed.

A drive is on to stop these destructive marauders, especially the jeep, dune buggies, and tote goats. State laws are being proposed.

It is time to restrict motor vehicles to recognized roads.

As a corollary of saving our parkland and forest lands for those who do not travel by motor vehicle we should follow Oregon's example. In 1971 Oregon passed a law providing that "Footpaths and bicycle trails shall be established wherever a highway road or street is being constructed, reconstructed, or relocated." And provision is made for use of moneys out of the state highway fund for those purposes.

Under the Act of May 28, 1963,[6] Congress created the Land and Water Conservation Fund, administered by the Bureau of Outdoor Recreation in the Interior Department. Under that Act the federal government can match state funds in acquiring, for example, land for footpaths and land for bicycle trails. BOR is helping states round out their holdings of easements or fee interests along the Appalachian Trail (a national trailway)[7] and BOR is also helping finance bikeways, two recent ones being a nineteen-mile path along the ocean below Santa Monica, and the other a three-mile route along San Diego's shoreline. Critically needed are bike-lanes, say a six- or eight-foot section of an existing roadway, for bicycle use only. Then our people could get to work without being run over, and the quality of our urban air would greatly improve. People who ride a bike to work now chain it to a tree and remove one wheel, taking it to the office along with the briefcase. Bike parking lots have yet to be designed.

We need not be strangled by highways, nor need travel by motor vehicle be the compulsory way of transportation. Motors may be our slave; they need not be our master.

LAND USE

Son of man,
You cannot say, or guess, for you know only
A heap of broken images, where the sun beats,
And the dead tree gives no shelter, the cricket no relief,
And the dry stone no sound of water. Only
There is shadow under this red rock,
(Come in under the shadow of this red rock),
And I will show you something different from either
Your shadow at morning striding behind you
Or your shadow at evening rising to meet you;
I will show you fear in a handful of dust.

T. S. ELIOT,
The Waste Land

It is within the power of the legislature to determine that a community should be beautiful as well as healthy, spacious as well as clean, well-balanced as well as carefully patrolled.

Berman v. *Parker*,
348 U.S. 26, 33 (1954)

Most ecologists agree that if fire were kept out of the Pine Barrens altogether, the woods would eventually be dominated by a climax of black oaks, white oaks, chestnut oaks, scarlet oaks, and a lesser proportion of hickories and red maples . . . Fire favors the pine trees because they have thick bark that provides insulation from high temperatures, and also because burned ground is just about perfect for pine seedbeds.

JOHN MCPHEE,
The Pine Barrens

ZONING, ONCE SERIOUSLY CHALLENGED, is now the accepted norm. And it can be employed not only for health purposes but for urban esthetics as well.[1] Yet every zoning ordinance allows for variances, and they are widely used.

My experience indicates that the start of our loss of open space, bits of parkland, and islands of wilderness in urban areas is at the level of the zoning boards. Real estate and construction companies use local zoning boards to destroy bits of wilderness. Builders, contractors, and real estate men are the applicants asking zoning boards for "variances." They request permission to deviate from the norm—to crowd more houses on the standard plot, to receive waivers so that their septic tanks can be built closer to the stream or lake, to cut down trees that have been protected, to pour concrete in areas hitherto reserved for green plots. Across the country these pleas nibble away at the small sanctuaries we have left.

Courts and agencies are beginning to apply ecological standards to their rulings on zoning. A recent example is *Nattin Realty Co.* v. *Ludwig,* decided September 23, 1971, by the Supreme Court of New York. A complex of garden-type apartments was denied a permit in light of the local water supply and the proposed sewage disposal facilities. The court said: "If there is substantial evidence sustaining the municipality's determination to rezone because of ecology, the court should not void such legislative determination." Yet urban planning has been so dismal that there is a movement afoot to put New Yorkers and people from Los Angeles on the endangered species list.

Most rural areas are experiencing a galloping subdivision promotion. Developers promise golf courses, country clubs, many new customers, and general prosperity. They can easily get the local merchants on their side. This is one source of pressure on local zoning boards. Another is the dream of high-priced lots bringing in fantastic taxes. Developers buy rural land for $300 to $400 an acre and convert it into $10,000 to $20,000 acres.

Usually a third of the price is used for acquiring the land, laying out streets, providing a water supply, and so on. Another third is spent on advertising and sales. The last third is profit.

At the local level it is natural that realtors, merchants, and bankers do the actual planning. They usually plan for "development"—not for the maintenance of open space, nature trails, parks, and recreational areas. They plan that way not because they are venal men but because they see only today and tomorrow and do not have the long view that a regional planner should have.

Moreover, the passion for "development" possesses many communities. There are, however, increasing exceptions. Loudoun County, Virginia, recently rejected a plan that would have given it a new 13,000-person community, because the people preferred the rolling hills to the sewage pipes, the highways, the litter, and the congestion that follows mass movements of people.

It is unfortunate that at the local level any development at all is often deemed desirable. The threat today is more ominous than ever. The population pressure is real. So is the power of the promoters, who in the seventies are huge corporations. One such promoter, in 1970, had about 85,000 acres for "development" in California alone. The little agencies of local government do not stand up under the pressures which these powerful interests create. Statewide planning is needed to give overall instructive guidance to local agencies. At present we are barely at the threshold of its development.

California's Environmental Quality Study Council, in its 1970 report, predicted that if the state's "development" by real estate men continues without severe controls, California will become "a vast wasteland."

The experts say that before 1980, two million acres of new agricultural land must be equipped for irrigation in California and farmed to produce the food needed by the newcomers. Unless drastic steps are taken, three million acres of highly productive land, now devoted almost completely to irrigated agriculture, will be changed into city land. The battle lines are already forming between acquisition of the land for urban development and its retention for growing food. The planning needed requires an end to "speculative land practice" and the inauguration of an inventory of remaining lands on the basis of "highest use." And "highest use" does not necessarily mean highest economic use but the highest use environmentally speaking.

Oregon, to prevent the forced conversion of open-space land to more intensive use, enacted in 1971 a law that permits land to be assessed in the open-space category where it protects natural or scenic resources, or air, streams or water supplies, or promotes conservation of wetlands and beaches, or enhances the value of wildlife preserves or recreational opportunities and the like. In this way the assessor assumes that the highest and best use of the land is open-space use and does not assess at a higher valuation resulting in higher taxes. How the Oregon law will work is not known.

New York, with somewhat the same objective in mind, authorized in 1971 the formation of agricultural districts, and provided that the "portion of value" of the land in excess of "the agricultural value ceiling" shall not be subject to real property taxation.

California's experience with a like measure has not been a happy one. California has had numerous laws to encourage retention of lands for agricultural use and for maintenance of open space. The most important of these was the Williamson Act.[2] It was designed to preserve agricultural land and to relieve farmers of the increasing property tax. Under the Act, as buttressed by a later constitutional amendment, lands whose use is restricted by agreements under the Act are not appraised on the basis of their value in the land market but only on the basis of agricultural income derived from the land. The recent Nader report, *Power and Land in California,* calls the Williamson Act in large part a failure: (1) it was used by some as a tax haven to hold agricultural lands off the market and then at the end of a cycle sell them for large profits; (2) it had virtually no success in saving prime agricultural land from urban sprawl; (3) little prime land was put under the Act, most of the coverage extending to nonprime land located apart from the urbanizing area.

The Williamson Act was amended to include scenic highway corridors, wildlife habitats, salt ponds, managed wetland areas, submerged areas, and land for recreational use. The latter is private land open to use by the public for walking, hiking, picnicking, camping, swimming, boating, hunting, fishing and sports and games for which the owner charges a "reasonable" fee.[3] It is too early to know how the recreational aspects of the Williamson Act are working in practice.

Meanwhile, concrete is fast being poured for highways, and every year sees more and more prime agricultural land in California and elsewhere being lost. A recent California study of agricultural law in Ventura County shows that

—residential use of land brings in $40 million in revenue and costs $97 million in services
—industrial use brings in $8 million in revenue and costs $8.9 million
—commercial use brings in $2.5 million and costs $3.8 million
—maintaining the land in agricultural use brings in $16 million and costs $380,000 in services.

On that popular cost/benefit basis, the Jeffersonian idea that our people should be primarily agricultural is the ideal. But our orientation is now entirely different.

Is rural America to be a factory site or a place to live? Is the "yeoman dream" of the small farm an empty one? Earl Butz says, "Farming is not a way of life"; and he predicts that by 1980 another million farms will go out of business. Meanwhile we prepare our budgets, our education, our research and development for agri-business.

Zoning as a control device has broken down. Where there is a surge of population such as Napa Valley in California experienced from the San Francisco Bay area, zoning boards are likely to be overwhelmed both by pressures from landowners within the coveted preserves and from developers. The pressures will be to grant variances so that the unearned increment of value in the land may be pocketed. Local administration, centered as it is in realtors, bankers, and businessmen, is too weak a reed to protect the public interest. If the public interest is to be served, there must be some better insulation from the developers. That is why overall state planning of land use that puts an end to speculative ventures is the ultimate tool that must be used.

In 1970 Vermont announced a bold program for comprehensive land use. There is a state environmental board and nine district environmental commissions. Permits for the development or subdivision of land are required, and these may not be granted unless numerous findings are made, e.g., that the project will not result in "undue water or air pollution," will not cause an "unreasonable burden on an existing water supply," or "unreasonable soil erosion" or will not have "an adverse effect on the scenic or natural beauty of the area, esthetics, historic sites or rare and irreplaceable natural areas," etc. The board is assigned the task of preparing for the state a "land use plan" which will determine "in broad categories the proper use of the lands in the state whether for forestry, recreation, agriculture, or urban purposes." Development projects falling under the "land use" program cover "the construction of improvements for commercial, industrial or residential use above the elevation of 2,500 feet" but not "construction for farming, logging or forestry purposes" below that elevation.

Vermont acted because large corporations were planning

massive developments. The threat was partly to the ecology of the state, as where roads and buildings were planned in mountainous areas where the land was fragile and where an undisturbed soil was needed to maintain its water-holding capacity. The threat was also partly to small towns with small populations, low tax rates, and few municipal services.

For the first time, esthetics, recreation, and the other decencies of civilization become overall, statewide standards for land use, replacing the conventional standard of making the "fast buck," the force which has largely despoiled the American landscape.

Hawaii in 1961 adopted a comprehensive land-use program. Those who visit the Islands, however, get disquieting reports concerning its effectiveness in actual operation. Perhaps Vermont in time will be disappointed; but her start is auspicious.

In 1970, Maine created the Environmental Improvement Commission to determine whether the location of a commercial or industrial project at a particular site would adversely affect the environment.

The things that may adversely affect the environment are numerous.

A smokestack may foul the air of a vast valley.

A mill, diverting water for manufacturing processes, may be a menace to anadromous fish. Oregon recently charged that two mills on one of its rivers killed six million young salmon and steelhead as they started down to the ocean.

A curse that has been laid across Eastern lands by lumber companies has been their use of rivers to get logs down to mills. Innumerable dams have been built to regulate the flow, and the bottom has been filled with bark deposits and sunken pulpwood which render the rivers unsuitable for spawning. Moreover, decomposition of the wood reduces the oxygen in the water to levels critical to fish. Salmon, trout, shad, and alewives which once abounded in the Kennebec in Maine have disappeared, and people at last are up in arms.

Montana got the great Hungry Horse Dam and everyone cheered at the prospect of making marginal farmers into industrial workers. A huge aluminum plant came in and its fluoride gas began to damage coniferous trees. Soon it was discovered that the fluoride also infected the deer population, shortening their lives. Is not this too high a price to pay for industrialization?

Mercury intoxication has had a long history. The phrase "Mad Hatter" came from the day when hatters used mercury in the manufacture of felt hats. Poisoned by the mercury, their nervous systems disintegrated slowly and they developed symptoms similar to classic insanity. Today plants that make chlorine use mercury, which vaporizes sometimes at dangerous levels, affecting workers. Is not that too high a price to pay for chlorine?

While mercury is a notorious poisoner and occurs in natural form in the environment, technology has multiplied its use. Chlor-alkali plants emit much mercury into sewers and into the air. The pulp and paper industry does the same. Mercury is indeed present in paper products and liberated when they are burned. Mercury is used to treat agricultural seeds, which causes seed-eating gamebirds to become infected. Fish carry in their tissues five thousand times the concentration of mercury in the water. Nearly all the mercury in fish is lethal. Mercury poisoning of waters in this country has not been a serious threat only because we are not a nation of fish-eaters. Smokestacks discharge mercury, and there is evidence that mercury reaching the environment in that way exceeds mercury reaching watercourses through industrial wastes. There is mercury in most food baskets which the housewife brings home. We have barely started monitoring the infection of food by mercury.

Other industries may pollute the community with trace metals such as cadmium, vanadium, beryllium. The toxicity of trace metals is well established; a dozen or more are now suspect. Like mercury, they may be methylated by microorganisms in waterways to produce highly toxic substances that accumulate in fish and other sea products. As I have said, many of these trace metals were in man's customary environment. But such historical records as we have indicate that they were present only in minute amounts. After 1940, however, following a great technological surge in chemistry, these toxics began to reach dangerous levels. An intensive scientific search is on (1) to see what can be done about the vast tonnage of metals in our waterways and (2) to learn how to avoid future accumulations.

Cadmium is particularly suspect. Cadmium is used in various manufacturing processes and enters the atmosphere in that way. Cadmium is discharged in liquid wastes from industrial processes. How much is not yet known; but some watercourses, e.g., the Hudson River and Lake Erie are, as some scientists put it, a good place "to mine cadmium"; and there is a whopping amount of cadmium in some water taps. It also is in phosphate fertilizers; and in that way the chemical gets into crops. Pesticides often contain cadmium, which enters potatoes, rice, other vegetables, apples and other fruit, and tobacco. The storage of poisons is in the kidneys and the liver and medical studies now indicate that its main impact on humans is in an increase of hypertension and high blood pressure.

Though the problem of land use has world-wide implications, it starts with us at the zoning board level. What industries—what industrial processes do we want in our states? Who would like to live downwind from a nuclear energy plant?

The federal government is an accessory to the growing crisis of local governments. It has opted for mobile homes, rather than permanent shelters. The "parks" where they are located are often just over the edge of the line marking the jurisdiction of a local community. But they must hook onto electrical, water, and sewer lines. Their children need schools; the "parks" need police protection and garbage collection. The "parks" seldom pay taxes equal to the burdens they place on these local services.

Resort to trailers came about because so-called public housing costs more than most American families can afford. Trailers are a stopgap; but they seriously imperil the concept of wise land use.

Real estate developers have so crowded houses in scenic areas and septic tanks have so multiplied that after a few years the owners begin to realize they are poisoning themselves in their own wastes. This has happened on riverfront and lake properties and in coastal areas where people flock to get a piece of a river or a piece of a bay or ocean in their front yard. New Jersey has adopted strenuous measures against these practices.

Wise land use requires statewide planning, as emphasized by Senator Henry Jackson, original sponsor of a federal assistance program to such state projects. It comes as a shock that many of our prized agricultural lands have become industrial sites, when the plants could have been placed on outlying unproductive ridges. It comes as a shock to see whole fertile valleys jam-packed with houses, when a central urban unit such as Paolo Soleri is designing in Scottsdale, Arizona, would accommodate the needs of people and save for the entire community the use of the surrounding land for varied activities.

Some ask, Why allow planners to interfere with the introduction of a pulp mill? It will put 500 men to work, increase the consumers by 2,500 and be good for every shop up and down Main Street. That is the historic argument. But what about the odor? What about the sewage? What about the new schools? What about the new freeways that follow in the wake of new factories?

Air pollution is a reminder that our technological achievements—the automobile, the jet plane, and the power plants—are, in environmental terms, costly failures. Water pollution tells us that the waters' limited, natural self-purifying cycle has broken down under stress. Soil depletion is a reminder that organic matter, in the form of food, is being extracted from the soil at a rate that exceeds the rate of rebuilding of the soil's humus. It also tells us that while loading the soil with inorganic fertilizer is capable of restoring the crop yield, it increases our water pollution.

These three ecosystems have been polluted by synthetics such as pesticides, detergents, fertilizers, plastics, and radioactive substances, which the systems cannot convert or absorb, and which accumulate in places harmful to the ecosystems and to man. As Barry Commoner says, "There is something gravely wrong with the way man uses the natural resources available to him on the earth."[4]

Technology and the profit motive have carried us far down the road to disaster. It is indeed a desperate race to institute preventive controls that will save the ecosystem.

Some are willing to leave the problems to technology to cure; and that is known as the technological "fix." But the ultimate decisions involve value judgments. In what kind of a place do you want to live? What environment do you choose for your children? How close are you to a nature trail? Do you have nearby a touch of wilderness in a forest?

These are crucial decisions to be made by the people, not by the experts in technology.

The power of public opinion was illustrated in 1971 when Delaware banned new heavy industry and offshore facilities for transferring bulk cargoes such as oil along its hundred-mile coastline. Delaware, like many other governmental units, has opted for quality of living, not growth for growth's sake. In 1971 Maine followed suit by rejecting through her Environmental Improvement Commission a proposal to build offshore oil refineries. California, however, marches the other way. Coastal areas are extremely desirable to land developers; local communities looking for "dollar developments" are easy to handle. State control of coastal development would seem to be closer to the ideal. By late 1971

California seemed doomed. Bills that would have protected the coastal areas from wall-to-wall divisions were defeated in the state legislature, and developers gleefully launched their plans to despoil California's northern beaches as they already had despoiled the southern ones.

Public opposition to wall-to-wall divisions along waterfronts, or to airports, new mines, industrial waterfronts, and dams is called alarming and vicious by some; but it is a healthy sign that more and more of our people do not want industry for industry's sake or payroll for payroll's sake.

What is this idea called "progress" or "development" or Gross National Product? It deals mostly with things that can be measured in quantities—more to eat, more to drink, more to use. But what about the quality of life?

Do we want to ruin a mountain or a river or a lake "to produce the power to run a million electric toothbrushes, can openers, shoe polishers, and other such gadgets, to make a billion aluminum beer cans, to light a million neon signs in ten thousand honky-tonks?"[5] as Hal Borland puts it.

Those are the questions that must be answered; and leaving the answers to industry and technology will inevitably produce an ugly, scarred, and plundered planet. Unless we adopt ecological standards, our future "development" will make a "fast buck" for promoters and industrialists but ruin our valleys, despoil our rivers and lakes, and even plunder the hills that serve as a backdrop for people who deem quiet, serenity, and beauty the choicest luxury of all.

Our planning units traditionally have been the states. What plans do we have? Are we destined to be managed and regimented by real estate brokers?

Our next eyesore promises to be Alaska. Once the claims of natives to land are resolved (as they are about to be) the remaining public lands will be open to appropriation and exploitation by individual and corporate interests. No comprehensive land-use plan for Alaska has been designed. The "development" as of 1972 promises to be as exploitive, haphazard and destructive as the opening of the West was, over a hundred years ago. History appears to teach no lesson.

Each state of course has some machinery that could be used to effectuate planning. Vermont, as I have said, has the most comprehensive land-use plan of all the states. Overall, Oregon has probably the most pervasive conservation measures, though they fall short of planning.

Oregon's smoke-control law is strict and severely enforced.

Oregon's water-pollution laws have made even the much-abused Willamette River swimmable.

Oregon, in a plan similar to the California law I have mentioned, has its new Open Space Tax Law, which shows promise.

In 1970 Oregon enacted a Scenic Rivers Act so that at least some of its pristine waters will be forever free-running rivers.

Oregon in 1971 arranged for footpaths and bicycle paths to be built alongside the highways that are constructed.

What about our waterways? As noted, Oregon has made some of its sparkling rivers scenic waterways with a sanitary corridor on each bank. Maine did the same with its famous Allagash. Washington, Tennessee, and Kentucky have been struggling to do the same with pieces of their pristine watercourses. Beginning in 1963 a joint Interior-Agriculture wild rivers study team has been considering some seventy rivers or river segments for classification as national rivers. Some of them have been made national rivers by Congress, the first being the Ozark National Scenic Riverways, formed out of portions of the Current River and the Jacks Forks River in Missouri.[6] The most recent is the Buffalo River in Arkansas.[7] These state and federal programs for making sanctuaries out of waterways are minuscule as compared to the destructive practices of the Corps of Engineers, TVA, the Bureau of Reclamation, and the Soil Conservation Service, already discussed. But the scenic river approach is the key to preserving wildlife, picnic grounds, swimming holes, fishing and hunting areas, and hiking trails that reflect the dominant amenities in our way of life.

Over a period of many years Oregon has through the action of her various branches of government kept in the public domain all the beach land between low and high tides. Of our total shoreline of our forty-eight "lower states" barely 2 per cent is in public ownership. Of the amount suitable for recreation only 5.7 per cent is available to the public. California is one of the worst offenders. Of her 1,272 miles of recreational shoreline, 1,023 are privately owned. Of 649 miles in Massachusetts, 631 are privately owned. Nearly all of Maine's are privately owned—2,578 miles out of 2,612 miles. And all but two of Louisiana's 1,076 miles are in private hands. In Florida, out of 2,655 miles, 2,372 are privately owned.

Senator Jackson has sponsored bills which would authorize acquisition of easements for public access to beaches, the funds to come from the land and water conservation fund on a 25 per cent state and 75 per cent federal matching base.

I have spoken about urban lands. But the health of a city depends, of course, on the health of the inner city, the ghettos where nearly two-thirds of all nonwhite families live. It also depends on the health of the fields and forests that are nearby. For those fields and forests have a direct impact on the water supply of the city, on its oxygen supply, and on the opportunities afforded its people for leisure activities.

Those lands more often than not are private, not public. In fact, over 90 per cent of our forest lands are in private ownership. Conservation laws should place responsibilities on owners of *private* forest lands both to protect and properly to manage their forests. Those acres should be under sustained lumber-yield management. Under such a plan raids on the federal sanctuaries to keep us supplied with lumber would be wholly unnecessary.

There would be many conservation dividends if private

owners were required to adopt conservation standards. People in the East—where there are few national forests—know this too well.

The method of clearing hundreds of acres of private lands for construction projects and the logging practices on private lands have made our Eastern rivers run red with silt.

California in its Forest Practice Act[8] established controls over logging of private lands. But the control was placed in committees composed exclusively of persons financially interested in the timber industry, i.e., timber owners and operators. While a state board must approve or disapprove the forest practice rules proposed by the industry, it had no authority to impose rules that would insure reasonable environmental and public protection against logging abuses. The rules proposed by the industry had the force of law. Yet the Act contained no guides or standards to prevent the abuses at which the Act was aimed.

The court, in *Bayside Timber* v. *Board,*[9] ruled on September 16, 1971, "Truly fundamental issues should be resolved by the legislature"; and their resolution cannot be delegated. "It is most certainly a truly fundamental issue" whether "the states' environment and ecology, and the public generally are to be protected by law from harmful practices of the logging and timber industry." The court, therefore, held that the Act in those respects was unconstitutional. And that holding was in the American tradition that lawmaking is a public function and cannot be delegated to private groups. *(Schechter Corp.* v. *United States.)*[10]

Urban planning goes much further than the city limits. It will often entail developing parks or wilderness areas on the perimeter of a city or nearby. Stretches of woods that are intact should of course be preserved. But many areas surrounding cities have already been denuded. It is, however, possible for the city, state, or federal government to acquire cut-over, eroded, and logged land and set it aside as a preserve. Mother Nature works wonders with unmolested tracts. Sunshine and water perform miracles.

As a nation we are greatly short of woodlands in the form of parks and wilderness areas. They have not kept pace in number or in size with our population. That is true of the Far West; and it is doubly true of the East.

In 1901 James Wilson, Secretary of Agriculture, recommended the creation of natural forests in the wondrous Appalachia. President Roosevelt agreed and sent the proposal to Congress. But the land was all in private ownership, though it then could have been purchased for from two to five dollars an acre. The power of lumbermen and farmers was great and the will of the federal bureaucracy was weak. The Arcadia, Shenandoah, the Great Smokies and Blue Ridge National parks and the Allegheny, Green Mountain, and Monongahela national forests were in time established. But the rest of these unique and exciting lands were logged, gutted, or strip mined and left as Wilson predicted to "the caprice of private capital, which has no interest beyond the profits in the lumber industry."

It is time we started reclaiming what we have left of the wondrous Appalachia and Alleghenies. It is time we set aside in perpetuity the islands of high grass country we have left in the Middle West. Both East and West have much stumpage that could be acquired and put in preserves, where in another fifty or one hundred years great new forests will stand.

Wise land use cuts very deep. That resource is limited, as population pressures remind us. It behooves us to insist on stiff ecological standards for the management of the complex problems of air, water, soil, and open space. As already noted, Congress in the National Environmental Policy Act of 1969[11] directed each federal agency to include in every recommendation or report on proposals for legislation and other "major federal actions significantly affecting the quality of the human environment, a detailed statement by the responsible official on—

(i) the environmental impact of the proposed action,
(ii) any adverse environmental effects which cannot be avoided should the proposal be implemented,
(iii) alternatives to the proposed action,
(iv) the relationship between local short-term uses of man's environment and the maintenance and enhancement of long-term productivity, and,
(v) any irreversible and irretrievable commitments of resources which would be involved in the proposed action should it be implemented."

That would seem to be the minimum which each state should require of its agencies. It would also seem to be the minimum which zoning boards should require before issuing permits.

But these statements run the risk of being not meaningful but perfunctory. The one submitted by Interior on the Alaska pipeline was meaningful; but it was 3,550 pages in length, filled nine volumes, cost $42.50 per set, and had no index. Seven copies were available for inspection in the "lower forty-eight" states.

Some agencies have refused to submit such reports and have been brought to heel only by court orders. Some reports have been piecemeal, giving detailed summaries of individual projects but never the total impact of a series of related projects on an entire basin. Bonneville Power—the former knight in shining armor—now files one impact statement covering everything it plans to do in the next year but not disclosing the true impact of any single project. Some agencies seek to control dissent and withhold scientific documents that bear on the project but that are adverse to it. The tendency of agencies is to play hide-and-seek with these NEPA requirements so that there will be a better chance of having their own way.

Agencies that submit environmental impact statements do so grudgingly, and they seem to cater to "economic" justifications when they come to benefit:cost ratios. The economics of protecting or not protecting a virgin stand, an estuary, an alpine meadow, are not irrelevant. But the web of life that is jeopardized in most schemes requires vastly different considerations. Estuaries represent imponderables as well as dollars. Virgin stands of timber marching to a climax forest may be indispensable to moose, eagles, and ivory-billed woodpeckers.

The National Environmental Policy Act of 1969 has paid great dividends. The bureaucrats are forced to think in ecological terms; and the long-suppressed naturalist or biologist in the bureau is at last allowed to educate his superiors, his Congressman and Senator, and the public at large.

The experts say that with high-yield agriculture we could get along with no more than two-thirds the land now fenced in farms in the United States. That means that fences now around plots which farmers are paid to leave fallow could be removed and the land restored to open prairie, river bottom, or potholes.

It means, moreover, that we could take, say, the northern half of Minnesota and Michigan for national parks.

It means we could turn much of the Appalachians and southeast Ohio (without disturbing any city or village) into old Daniel Boone country.

It means we could revert a lot of the West, including the high grasslands already mentioned, to open lands where the deer, antelope, prairie dog, and coyote could run once more.

Secretary Rogers C. B. Morton of Interior in 1972 designated about three million acres of desert lands in California for recreational use—areas being chewed up by dune buggies and tote goats, being used as garbage dumps, and losing their archaeological artifacts to vandals. Interior will propose to Congress a multiple-use desert master plan. "What we really need," Morton said, "is to develop a desert ethic. If we fail to instigate a land-use system in America now, the next generation will live in an intolerable environment."

One of the most critical long-range problems of land use concerns a serious and conflicting federal policy. We pay farmers not to grow crops in states with high rainfall. Yet we subsidize farm irrigation in states with low rainfall. Arizona pumps from underground pools about 3.5 million acre feet of water a year (an acre foot is the amount it takes to cover an acre of land one foot deep, i.e., 326,000 gallons). That is the amount of water Arizona uses to irrigate lands for feed grains and animal forage. Those are low-value crops easily grown in high rainfall areas of the East. Why spend billions of federal funds to provide water to replenish Arizona's diminishing supply when wise land use would put an end to profligate water waste?

In terms of the national economy, the Central Arizona Project approved September 30, 1968, was probably a massive mistake. It sends Colorado River water to Arizona to grow crops that are not needed. The taxpayer pays double—first for the project and second for managing the surplus crops.

These are tremendous decisions, which only an informed people can make but which, I think, they will make wisely once they are freed from the powerful influence of the present public relations approach to conservation.

POLITICAL ACTION

It is the political order that is making fateful decisions about man's survival in an age haunted by the possibility of unlimited destruction.

SHELDON S. WOLIN,
Politics and Vision

There is only one revolution tolerable to all men, all societies, all political systems: Revolution by design and invention.

Every nation welcomed the transistor.

Every nation will welcome desalinization.

All the world, properly informed of the significance of the design and invention revolution, would welcome it.

Science, not politics, centralizes society. The telegraph wire communized the world.

BUCKMINSTER FULLER,
I Seem to be a Verb

Now the capriciousness of man altered the scheme and introduced a spreading degradation of nature.

Yet man possesses the ultimate complexity denied all other life: he is the world—its physical and energetic components—at last aware of itself. His intellect is capable of perceiving the change and blight he has brought to the earth, as it is capable of halting and restoring some of the loss. He is aware there are too many of his kind; that by his very numbers he burdens the sky, earth, and its waters. As a conscious manifestation of the world's evolutionary progress, he is bound to the creation of new levels of complexity for survival.

WILLIAM H. AMOS,
The Infinite River

A PRIVATE CITIZEN has no Pollution FBI to call upon when toxic pulp and paper mill wastes move across state lines in the air or by the tidal action of interstate rivers. He has no Consumer Zoning Commission to put all power lines and telephone lines underground in order to preserve the beauty and serenity of urban and even rural areas. New York took the lead in 1971 in proposing that all existing overhead local service lines be placed underground, but Consumer Zoning Commissions are mostly lacking. The consumer has no Pollution Securities and Exchange Commission to see why each family is defrauded each year by assessments for air and water pollution controls that never come; he has no Pollution Food and Drug Administration to assure him the air he breathes and the water he drinks are pure. The consumer needs these protections; yet it seems from reading the reports of Congress in 1970 and 1971 that all levels of government have failed him.

The place to start is with the head of each agency dealing with the environment. Many of them, inlcuding the heads of the Forest Service, the Park Service, the Bureau of Land Management, the Fish and Wildlife Service, the Corps of Engineers, the Soil Conservation Service,the Bureau of Reclamation, are promoted from within the bureaucracy or named by a Cabinet officer. They are entrusted with such broad authority that they can pour their own predilections and prejudices into such vague standards as "the public interest." They therefore should be named by the President and subject to Senate confirmation. Their philosophy and attitude and past affiliations can be fully probed and their fitness for the post can be laid alongside the critical environmental problems of the day.

To date, most federal agencies dealing with environmental problems make their decisions with no opportunity for the consumers, i.e., the public, to speak. The consumers are not entitled to be heard, to present evidence against a project, or to appeal from an adverse agency action. This is true of the Forest Service, the National Park Service, the Department of Transportation.

The public is not entitled to a hearing before the Forest Service begins to cut a virgin tract or before it pierces a wilderness with a new road.

The public is not entitled to a hearing before the Park Service launches a "development" plan that builds a village inside the sacred precincts of a park.

The public is not entitled to be heard before the Bureau of Public Roads designs a freeway and draws the lines of its location.

These agencies are indeed appalled at the very thought of consumers having such leverage over their actions. They say that an agency could not get its work done if it had to put everything down for a public hearing. *But important things —major items—should be aired publicly.* All bureaucracies —U. S., British, Russian—get inbred and dictatorial. Public hearings can create a stream of clear fresh air blowing away the cobwebs.

This raises the problem of "standing." "Standing" to sue has usually been considered to be a dry, technical problem. But in the field of environmental questions it has broad philosophical connotations. When man speaks against destruction of a valley, say by the Soil Conservation Service, he speaks for the entire community that lives there. The water ouzel is in that community; the red fox, the coyote, the bear are others; the deer is included; so are the birds overhead, the earthworms and golden-mantled ground squirrels underneath, the fish in the stream, and the rest of the wildlife dependent on the river. The entire community includes of course the trees, the shrubs, the spring beauty and all of the wonders underfoot. These members of the community have ways of communicating with each other. Even the coyote and the fox talk, though man does not understand what they say.

The members of the community, in other words, are in no position to be heard at any public hearing. Only man can speak for them. He who knows them and understands them and appreciates their role in the community has "standing" to speak before they are destroyed. That is the essence of the "legal" question in the environmental setting. If ecological standards are to control, as Congress has decreed in the National Environmental Policy Act of 1969, then the place of all these phases of life along the river must be weighed and appraised. It would be an unnatural, senseless thing to destroy them, certainly without a hearing. And the hearing, to be meaningful, would have to operate on the assumption that all life is sanctified and that its sacrifice should not be tolerated except in case of a clear overriding social need.

In this connection the most important advance was a Michigan Act of 1970, giving any person standing to sue any branch of government in Michigan or any industry "for the protection of the air, water and other natural resources and the public trust therein from pollution, impairment or destruction." Senator George McGovern has been pushing a like federal bill for the protection of the air, water, land, or public trust of the United States from "unreasonable pollution" resulting from an activity "which affects interstate commerce."

Congressman Udall is sponsoring a bill that gives citizens standing (1) to present evidence and arguments before federal administrative agencies; (2) to challenge administrative decisions in federal courts; and (3) to sue in federal courts to protect the environment.

So far as air pollution is concerned, the 1970 Act is replete with opportunities for private persons and groups to participate in hearings. And it allows "any person" to sue civilly even the United States government for violation of emission standards and federal or state air-pollution orders, save where the federal or state government is itself seeking to enforce the standard or order in question.

A word of caution is necessary. Resort to courts is not the ready answer, though it is the impulsive thing to do. Courts, however, have no great experience on the environmental front. Legal precedents, moreover, usually were established in days of laissez faire when equity normally acted to protect business, not the consumers. Laws and rules change; and courts are responsive. But those who go to court are met by batteries of industrial lawyers. Time and expense are on the side of status quo. If there is to be progress on the environmental front, there must be political action at the state and federal legislative level. There must be political action at the federal and local agency levels, in order to get new rules and regulations enacted. There must, moreover, be persistent, informed participation in agency proceedings.

More basic changes are needed. One who sits in Washington, D. C., for long begins to wonder what administrative "expertise" really is—the talismanic quality which is often given great credence. It often is only the accustomed way of doing things. In fact the accustomed way may be in need of uprooting and need to be entirely displaced. Across the world, more often than not, doubts arise as to the wisdom of letting the "experts" decide what to do. The "experts" need to rejoin the battalions of workers to get new perspectives on their jobs. A judge, for example, who had a regular tour cleaning toilets and cells in a prison would come to the awful task of judging with greater humility. The factory manager who joined the brigade of workers every fortnight would be certain to gain new insights and perspectives.

The other side of the coin is worker participation in the making of production policies of the factory. The reports out of Peking indicate that since 1966 workers in factories do help make policy decisions, all to the benefit of production records. The custom is neither Chinese nor American, though we greatly honored it in our town hall traditions. *Let the people be heard* is the motto. Let them be heard not only in casting ballots but in debating municipal or factory or wood-lot policies. Let them, in other words, help shape the policies on which the fate or future of the community depends.

As indicated, all bureaucracies are paralyzing and suffocating. Ours are bad enough; the Russian and the Italian and the Iranian and the Greek are insufferable. They are closed societies responding to some special interest or reflecting one prejudice, one point of view. They need to be jolted, aroused, and alerted to a new and different problem. Only the voice of the people speaking in unrestrained but measured voices can rectify the situation. "Let the people participate in policy decisions" should be the rule, not the exception. The people know, far better than the experts, what poisons are permissible west of the Mississippi and what slopes should be barred forever from clear-cutting.

Nuclear energy is a technical matter and a mysterious one too. Only the "experts" understand the process. And so the public is trained to stay in the back seat and let the "experts" take control. In the case of AEC that is a tragedy. AEC is not composed of "experts"—one member is an economist, two are lawyers, one is a biochemist, and one is a mechanical engineer. That is no criticism of the Agency; its diverse composition is indeed a virtue. But the composition of AEC shows that it is indeed only a "lay" group. Informed, intelligent laymen drawn from the community and advised by radiologists can speak with the same authority as AEC members.

The tragedy of keeping the people out of policy decisions is compounded by the *secrecy of agencies*. How does one find out the strontium 90 content of a plot of land in Hawaii, Duluth, or Miami? He cannot find out, even though that knowledge may be critical as to whether he plants vegetables or grazes cattle.

What is the location and design of the highway? What environmental damage will it do?

What are the ecological hazards involved if a Corps of Engineers dumping permit or dredging permit is granted for a particular estuary? Will the permits issued by the Corps be in substance permits to pollute? Why should not the data on which the agency proposes to act be made public so that consumers, if need be, may summon their opposing experts?

The Corps of Engineers, once addicted to almost *pro forma* hearings, changed its rules in 1970 and allowed the public easier access to hearings and provided more meaningful hearings.

In May 1972 the Food and Drug Administration decided to make available to the public about 90 per cent of its documents heretofore classified as "confidential." At long last there will be the full glare of publicity on drug applications, food additives and the like. Perhaps this turnabout will in time disclose how many cattle and sheep are being contaminated by the synthetic hormone DES, which is used to fatten animals and which, it is charged, is a cause of cancer.

A classic example of how government agencies work hand-in-glove with industry, keeping secret their plans, is illustrated by the development of the blueprint and legal groundwork for the six power plants in the Arizona, Utah, and Colorado area already mentioned. The Department of the Interior and its Bureau of Reclamation worked quietly and silently with the power consortium that will run the power plants. They did all the planning, testing, negotiating with the Indians, the leasing and the contracting with no publicity whatever. Not even the National Park Service knew of them. Neither it nor any of the state conservation agencies, greatly affected by the project, had a chance to be heard. There were no public hearings, no hearings before Congress, no chance for conservationists to ask questions or make studies, no opportunity for objective studies of the impact of the project on air, water, or quality of the environment, no reports to the Federal Power Commission. When the word finally did get out, everything was signed and delivered and the work was starting. The reason for secrecy, like the cause of anger once the secrecy was discovered, is obvious: the project, if aired and publicly discussed, would never have been adopted, because it promises to be ruinous to the beautiful, the unique, and the fragile Southwest.

A few of us—including students and faculty—were having a discussion at the University of West Virginia following an evening of ecological questions and answers. A professor, wise and experienced in air and water pollution, spoke up. He mentioned the few states doing a good job, but talked mostly about the failure of the other states. I asked him what he thought made the difference between effective and lackadaisical action. His reply was revealing: "When corporate officers are sent to prison for not complying with pollution laws, the air and waters will become clean rather quickly."

That problem is not peculiar to America. Pollution of rivers or lakes in Russia is punishable by corrective labor for a period of up to one year or a fine up to three hundred rubles. But industrial wastes are rampant; and 60 per cent of all Russian sewage enters the waters in raw form. Fines are levied against the enterprise involved; but the fines are used for civic improvements, which lead local governments to look on pollution indulgently. Fines are seldom levied on managers of the enterprises, though dismissal and sentencing are increasingly used in aggravated cases.

In Russia plants are opened even though they do not meet pure water standards and even though opening under those conditions violates the law. The pressure is on in Russia, as in this country, to keep violators in business and new plants open. The reason for dumping the wastes from pulp mills into Lake Baikal, the purest of all fresh-water lakes, was Russia's need for pulp.

Whatever public relations agents may say, in Russia there is no paramount concern for the environment. Such public concern as exists is subordinate to the engineering or technological concern for doing an efficient job. Russia perhaps does not yet realize that that turning the problems of the environment over to the technocrats is almost certain disaster. Russia promises to be an easy victim, for she has no First Amendment rights which conservationists can use for speaking, pamphleteering, and picketing.

As we have seen, industry is always looking for means of avoiding or even evading environmental regulations. In our country, one of their techniques is to control the agency entrusted with protection of the public interest. This influence can be venal and sometimes is. The identity of interests between the regulator and the polluter may be more subtle and not corrupt. It may be found in a common ideology that laissez faire is better than government control. It may reflect a silent hope that in time the agency staff will be working for the more affluent polluter. Or the voice that should be protecting the public may have counter pulls that silence it.

Man-made radiation emitted from nuclear power plants, now spreading like mushrooms across the country, presents an ominous specter. This sightless, odorless, tasteless killer of life is a silent and mysterious invader. Most of us are not sufficiently sophisticated to argue the case pro or con. Only the scientists are qualified. But many of the scientists are on the payroll of the promoters of these devices or on the payroll of the AEC or the Pentagon, who sponsor the promoters. Many remain silent because they fear reprisals if they speak out or probe the critical issues of the day or espouse, at least publicly, human values. There is a point of pollution by radiation which puts man beyond recall. Drs. Tamplin and Gofman discuss the problem in their frightening book, *Population Control Through Nuclear Pollution*, and urge their fellow scientists to "end the cruel illusion that if environmental deterioration threatens, science will come to the rescue."

There are degrees of difference when we turn to pesticides,

strip mining, the use of inland waters and oceans as dumping grounds, the infection of percolating waters, and the violent changes in ecology caused by pipelines carrying hot oil over and through the icy tundras of Alaska. Whatever the problem, it is essential that the voice speaking for humanity be strong, for the polluters are about to possess the earth.

We are crippled by the historical fact that for decades we have intrusted conservation to the "outdoorsmen"—the hunters and fishermen—plus the nature lovers. Almost anything that did not damage hunting and fishing was promoted in the name of conservation. The public relations men for the vested interests who wanted to exploit the out of doors got the message. The American theme became "wise use," the inference being that if we did not "use" a wilderness, we were not very wise. This idea, hitched to the notion that the aim of life was to convert the wonders of nature into the dollar, led us to abandon the idea that some things—some wild areas—some pure waters—some living, pristine estuaries—were worth saving for their own sakes, whether anyone "used" them or not.

The National Industrial Pollution Control Council (NIPCC)—composed of industrial representatives—advises federal agencies on policies to pursue and policies not to pursue. Its audiences with federal officials are private, not public. Its members are real policy-makers, having an inside track that no ecologist or conservationist enjoys. These policy decisions are political decisions, involving what the cost of environmental protection will be and who will pay it—taxpayers, consumers, or industry? If all three, in what amount? Industry sits at the table where those decisions are made. Environmentalists want room at that same table. If the ever-present advisory committees representing corporate interests included the unrepresented environmentalists, they might become "a desirable avenue for public access to government decision makers."[1]

A March 1972 governmental report, for example, made

headlines across the country by predicting great economic loss to manufacturers who are forced to comply with anti-pollution legislation. It raised the specter of higher sales cost and cheaper import competition; and of shutdowns of small factories, local recessions, and unemployment. And it neglected to figure any possible economic benefits deriving from cleaner air and water. But it did make clear that of the 12,000 plants studied, about 800 would close anyway between 1972 and 1976; and while pollution abatement costs would cause some 200 more to close, they were vulnerable for other reasons and would likely have closed anyway a few years later. This report, sponsored by the Environmental Protection Agency, the Council on Environmental Quality, and the Commerce Department, was really farmed out to eleven private consulting firms—and the NIPCC.

If the voice of reform comes, it will come from the people, not the agencies. Only an active, enlightened citizenry can keep the bureaucracy energized and on the straight and narrow path.

We need what Michael Frome has called "patriots without a price tag."[2]

If the profit motive is our sole standard, the environmental cause will be lost. Every control will cost money.

Corporations, as well as the federal government, state agencies, and municipalities, have assumed it is their God-given right to dump their wastes and sewage into rivers, lakes, and harbors. Is the voice of the people strong enough to take these perquisites away from entrenched interests?

Some argue that the decision makers must be well versed in technology if they are to act wisely. Under that theory EPA and CEQ (Council on Environmental Quality) would be headed by technologists. That, however, is a short-range and limited view. We deal with value judgments. A technologist may be able to tell us what it will cost to have swimmable water. But only the people can determine whether we should have swimmable water. The fact that the decision makers are "technologically illiterate" does not mean that they are unwise. The great decisions concern questions like "Do we want the earth and its air and its waters to be once more pristine pure?" The technologists play only a secondary role. They do not answer that question; they simply tell us how to achieve that goal.

Russell E. Train, Chairman of the Council on Environmental Quality, recently estimated that the cost to industry and government of air- and water-pollution control and the disposal of solid wastes would be about $105 billion over the six-year period 1970 to 1975. Yet even that high amount would be less than one per cent of our gross national product.

Congressman Charles A. Vanik of Ohio said, "There are and there have been sufficient local laws and authority to restrain every form of pollution abuse."

Speaking of sick Lake Erie, he added: "Local governments have vacillated, weighing pollution control enforcement against the industrial threat of plant shutdown with resulting unemployment. If Cleveland attacked a polluting industry, it could always threaten to move to another place."

Take, for example, the costs which some marginal producers would have to pay to meet the states' water quality standards. What if the higher costs will drive a marginal employer out of business? The issue is being raised in countless raw and gritty cases. For example, does a town prefer to turn its river into a sewer and employ two thousand men or forgo that new enterprise and have a river that is pristine pure and good for fishing, boating, and swimming? The answer of ecology is clear: a pristine river is the top priority. But the answer of chambers of commerce is often the opposite. Low-income workers—both white and black—are likely to think the same way. Moreover, riddance of rats in the ghettos is more important to them than saving the ivory-billed woodpecker in the Big Thicket.

One who picks up the ecology standard will not march long before he realizes that conservation cannot be a success if it seems irrelevant to the needs of the millions who live in squalor. When we speak of the survival of man and his habitat, we must think not only about wilderness areas, pure air and water, free-flowing rivers, abundant wildlife, and the like but also of ghettos, poverty, unemployment and the large bloc of people who suffer from pollution and its related ills and who are not beneficiaries of the affluence that produced it.

It is already being said that limitations on the increase of electricity would be harmful to the poor and the young. Only 8 per cent of nonwhites now own individual air conditioners, it is reported, as compared with 40 per cent of the whites; only 3 per cent of the nonwhites have dishwashers as compared with 12 per cent of the whites; only 41 per cent have vacuum cleaners as compared with 81 per cent of the whites.

We must, moreover, think of the workers and their unions, who today are often bewildered by the "job scare" which a corporation creates when it says it will "close down" before it meets a new air- or water-pollution standard. Civic action by ecologists and workers alike can sort out the bugaboos from the serious threats. Conservationists must be prepared (1) to insist that all industries be held to the same high technological standard so that there are no "pollution shelters" for any company; (2) to back legislation providing government compensation for a term of months or even a year or more for workers who lost jobs because of environmental controls; and (3) to throw their support behind job transition bills or public works job bills that would put unemployed people to work on environmentally useful projects.

The Environmental Protection Agency and the Department of Labor are doing something about this problem. When EPA believes that its pollution-control actions may cause layoffs in factories or plant closings, it advises the Labor Department, which starts setting up job-retraining programs and other projects to help the affected workers.

Many marginal enterprises will not be able to carry the

extra costs, though sturdier ones will be able to do so. Hence some suggest that all factories be required to pay a high fee for their pollution. The sulphur content of fuels could be taxed, a levy could be made of say ten cents a pound of pollution, as measured by the biochemical oxygen demand that effluents place on the waterways, etc. This would, in substance, be a license to pollute and the cost would be passed on to consumers who, it is argued, would tend to shift to manufacturers that had a lower environmental cost. Is that really a way to provide an incentive for industry to become pollution-free?

This is the stuff out of which great public debates will be generated.

Moreover, workers who are put out of work when EPA temporarily closes down a plant because of its air pollution bitterly complain that industry, not workers, should pay the cost. Technology, they say, is in the hands of the owners, not the unions.

This too is the stuff out of which the debates will be generated.

The age-long remedy for a new evil has been the creation of a new agency. These agencies have proliferated at all levels. Commonly they move with zeal for a few years and then succumb to inaction or to painfully slow action. They become oriented to the view of the group that they are supposed to regulate.

People curse the big interests and say that industry and finance have become stronger than government. Many ask for nationalization so that the people will have control of an industry.

Yet that new control would be through new agencies; and, people being what they are, the agencies under a socialist regime would be as inept and as paralyzingly slow and basically as timid as agencies under the regime of free enterprise. Perhaps more so.

The creation of new agencies is not the answer. The sad

story of TVA and its near-claim to being our Public Enemy Number One in the ecological field is proof enough. If there is to be a task-force approach to the problems of ecology, the people must be given incentives to get their consumer projects launched. They need incentives lacking today.

Organized conservation groups have acute problems. In order to obtain funds they need a tax exemption under the Internal Revenue Code so that contributions received are deductible by the donor. To maintain a tax exemption under the 1969 Act the group must not engage in lobbying either directly with legislators or at the grass roots. It may, however, give technical advice or assistance to a government agency in response to a written request or make available to a legislative body the results of nonpartisan analysis, study, or research. The result often is that only one side of an issue involving conservation may be presented to a legislature. If there is a bill before Congress to make a designated area into a national park, those opposed have no tax exemption problem. Steel mills that want to move in, or mining interests that want to maintain their status quo, or construction companies interested in logging—each of these can expend money fighting the proposal and deduct the costs as "necessary and reasonable" business expense for tax purposes. On the other hand, a conservation group dependent for its contributions on its tax-exempt status risks its tax exemption if it undertakes to give the federal or state legislature a true picture of the national values involved in making a designated area into a national park, though it may present broad social, economic and like problems of the type the government could be expected to deal with ultimately.[3]

Revision of these tax exemption laws so as to give conservationists an equal footing with those who would destroy a sand dune or a river or a sanctuary or a high ridge for the Almighty Dollar would enhance the environmental case.

Some private groups without tax exemptions are waging powerful political action programs on the environmental front. The points of danger, the conditions of crises are almost without number.

There is some response. Civic-minded people interested in correcting local environmental problems often produce spectacular results. Ten years ago, beautiful Lake Washington in Seattle seemed doomed by sewage. A few people were the nucleus of a promotional group that formed a new entity (Metro) which brought together some eighteen governmental units, built new sewage plants and interceptor lines, and ended the dumping of raw sewage in the Lake and into the Sound. All this was done with only minuscule federal aid under exclusively local management and ingenuity.

A like effort succeeded in saving beautiful San Diego Bay from becoming a dangerously polluted and largely useless body of water, so far as recreation is concerned. The planning was entirely local and the federal agencies played only a minor role in the financing.

Under Governor Linwood Holton, all of Virginia's rivers (except the Potomac) will be once more swimmable by 1974.

In Chicago a committee known as Campaign Against Pollution (CAP) has moved into action with unorthodox procedures against large industrial polluters. It made such a good showing in a stock proxy fight as to induce the company to reduce its use of high-sulphur fuels. To stop another large company from air pollution it moved against the directors. If one director was associated with a retailer, CAP picketed that retailer. If a director was running for office in a church, they moved in against him in the church election. CAP studies the tax assessments of a polluter and if it finds the company underassessed, it enlists the aid of civic groups (school boards, teachers, park districts, and the like) for an increase in the assessments.

Down in Jacksonville, Florida, the person who turned the tide against air pollution was Mrs. Lee Adams, who was first ignored and then ridiculed. One Florida Board of Health official told her to go home and learn how to play bridge. When officials found out she was serious, she started receiving threats. She announced that she would address the Junior League of Jacksonville on the subject of air pollution. That was a touchy subject because sulphuric acid fumes filled the air due to the use of high sulphur-content fuel oil by industry and by the Public Authority that generates electricity. Natural gas was the obvious answer. But the powers-that-be were against it. They had indeed got the Corps committed to building a deep-water canal for the purpose of bringing fuel oil into the city. The authorities were proud of their canal. The Public Authority found high sulphur fuel oil a money-saver. If it shifted to natural gas, it would have to reverse itself on a promise to consumers to reduce electric rates.

So before Mrs. Adams addressed the Junior League, an officer of the Public Authority called and told her she would be well advised not to make a speech on air pollution. She said she was going to. He said, "Well, if you must give your talk, don't mention natural gas." But she did. The Junior League supported her and the League of Women Voters supported her, and pretty soon she was no longer alone—but leading a civic reform program.

Verna Mize, housewife and secretary, has as her cause the "saving" of Lake Superior, that once was famous for its clear water. It now has a green-gray cast caused by the 67,000 tons of tailings of a taconite mine which are daily dumped in the lake. Mrs. Mize is known in Washington, D. C., as a "one-woman truth squad" who makes many trips to the Hill to obtain converts. Some experts disagree with her diagnosis of Lake Superior; but undaunted, she goes on with her campaign, not missing a day since 1967 in speaking up for her lake.

A citizen's group led by Marty Kent Jones led the protest against the Corps' destruction of Tamalpais Creek, Kentfield, Marin County, California, by channelization. Some were fined and sentenced to jail for their protests. But as

Marty Kent Jones said at the time, "A good cause grows faster than the wind; it exhilarates and is joyful in the face of obstacles." As respects the arrests of the protesters, she added, "Beneath the cause there is a love of nature; there is concern for humanity; there is brotherhood in action. This is creation, not destruction, and it drives people to take a stand and be counted."

Marty Kent Jones lost that battle. But she stated the philosophy that will in time turn the tide and enable us to win all the other battles.

Some architects are attacking "the visual pollution of the cityscape," as David B. Singer puts it, and they emphasize that the National Environmental Policy Act of 1969 shows concern for "esthetically and culturally pleasing surroundings." The Design Review Board of Riverside, California, was created, and the sentiment is that it is producing an improved-looking city at no appreciable cost increment. We may end up not only with clean and healthy but with lovely looking communities.

There is much publicity about the ecological standards being written into federal laws. One example is the airport law governing the Department of Transportation:[4]

"It is hereby declared to be national policy that airport development projects authorized pursuant to this part shall provide for the protection and enhancement of the natural resources and the quality of environment of the Nation. In implementing this policy, the Secretary shall consult with the Secretaries of the Interior and Health, Education, and Welfare with regard to the effect that such project may have on natural resources including, but not limited to, fish and wildlife, natural, scenic, and recreation assets, water and air quality, and other factors affecting the environment, and shall authorize no project found to have adverse effect *unless the Secretary shall render a finding, in writing,* following a full and complete review which shall be a matter of public record, *that no feasible and prudent alternative exists and that all possible steps have been taken to minimize such adverse effect.*" (Italics added.)

If read closely, it means that the airport wins and ecology loses.

Can a law be passed where ecology wins?

Can an Ecological Code of Ethics be drafted which will become effective?

Can saving the environment become as holy a cause as making a fast buck?

Religion can make a difference. The Hindu philosophy that even plants in a garden are kin to man has had revealing results. It has been discovered that certain kinds of music make plants grow, and other kinds retard their growth. Plants respond, in other words, as humans do, to music.

Trout, bass, ducks, have their spokesmen. Recycling paper is a cause for many. A dirty local smokestack is a rallying point for some. More and more people have generalized interests in alpine meadows, canoe waters, virgin stands of timber, pristine rivers, peaks and ridges. Those who speak for them speak for the entire ecological community that is represented there. Bears, coyotes, eagles, fish and perhaps even the spring beauty can and do communicate. But men do not understand what they say. When wildlife has a confrontation with a bulldozer, it can only flee; and the botanical wonders can only submit and be plowed under or die from spraying. Man, however, can speak for the entire ecological community.

In 1620 William Bradford wrote of the Puritan attitude toward the "hideous and desolate wilderness, full of wild beasts and wild men." And he added, ". . . which way soever they turned their eyes (save upward to the heavens) they could have little solace or content in respect of any outward objects."

That attitude is the historical animosity we have had toward the wilderness. We are now more conscious of the web of life in the total environment; and man therefore must fulfill his destiny of becoming not the exploiter but the guardian of all members of the ecological system.

Senator Alan Cranston of California has proposed the establishment of a 35,000-acre Pupfish National Monument. The pupfish is one inch long and is too small for man to eat, and is not fit for a home aquarium. Yet it is in danger of extinction. As Michael Frome says:[5]

"Still, I agree with Senator Cranston that saving the pupfish would symbolize our appreciation of diversity in God's tired old biosphere, the qualities which hold it together and the interaction of life forms. When fishermen rise up united to save the pupfish they can save the world as well."

There is a school or cult that extols "behavioral technology,"[6] which in terms of environmental problems means that man will refrain from polluting not on reasoned grounds but because he has been conditioned to want what serves group interests. If that is true, relatively few will decide what is best for people and use the levers of power that are available. That is why pessimists predict that the existing bureaucracy will so condition the public mind as to make the continuing destruction of the environment both "profitable" and "palatable."

The religion of Zoroaster—established in Persia about 1000 B.C. or earlier—had an ecological ethic. This ethic taught soil conservation, wise grazing practices, protection of timber lands against deforestation. Zoroaster taught that the entire material world is holy—mountains, lakes, the crops, earth, sky, wind, and waterways. Like man, the ox has a soul that cries out for justice; man and animals are inseparably linked, the sufferings of one are the sufferings of the other. Zoroaster taught that God created man specially to appreciate beauty, to love and respect nature—that man needs wholeness with nature to partake of immortality. The Iranian Adam and Eve appeared first in the form of a rhubarb plant (whose leaves are poisonous and whose stalk is edible) and then were transformed into human form. Their great Sin was in saying that the Destructive Spirit, not the Creator, brought into being water, the earth, plants, and other wonders. For that they were eternally condemned. What would have happened to Persia had that philosophy remained dominant will never be known. The Moslems conquered the land in the eighth century and those who escaped the deluge were the Parsees, the distinguished bankers and financiers of modern Bombay.

Whether we can make a religion out of conservation or give that cause a messianic mission is the critical issue of this day. Without it, laws will be wholly inadequate.

We used a task force to unlock the secrets of nuclear energy in World War II. We used a task force in unraveling some of the problems of Outer Space. The task force was used by us in World War II to stop a potential world-wide typhus epidemic before it got out of Naples, Algiers, and the Suez. A task force was used in the 1960's to stop an epidemic of yellow fever in Ethiopia.

Moreover, we have gone in for collective action on many fronts—defense, public works, highways, dams, flood control, schools—all of which involved either very large investments or investments which would yield widespread public gains which private enterprise could not capture.

We have the genius to approach the environmental crisis with a task force. Do we have the resolve to do so?

Much is written about environmental "backlash"—that people are sick and tired of all the talk about polluters and the like, and want to get back to normal affairs. But the problem will not go away; it will indeed get worse if people turn their backs on it. Radiation, air pollution, water pollution, bulldozing the wilderness, covering the earth with concrete, killing our estuaries, filling the air with noise—these things not only dilute the quality of life but destroy life itself. Industry and municipalities alike are now feeling the "crunch" of regulation and none of them is happy with it.

Public retreat means disaster. Our destiny certainly is not squeezing the last possible dollar out of the biosphere before we see it wallow and finally succumb. The Three Hundred Year War need not finally end in the destruction of all the participants. That is the commitment of two men of character and intelligence now guiding the national program—Russell E. Train of the Council on Environmental Quality and William D. Ruckelshaus of the Environmental Protection Agency. We must help make them strong.

Congressman John P. Saylor of Pennsylvania says:

"The American people are light years ahead of Members of Congress as to an awareness of our Environmental decline. In Congress, we see the dotted 'I's' and the crossed 'T's' of legislation. You, the public, see the bulldozer slashing away at the wilderness; you live amidst urban sprawl; you taste and feel the strong pollutants; and you clasp your ears as the jet rattles your dinnerware.

"The public support is out there for a massive congressional drive against further erosion in the quality of our lives, if only all of us in the Congress will bite the hot bullet and respond."

To date, our efforts on the environmental front have been largely public relations gestures. Wherever we turn—whether it be to air, water, radiation, strip mining, estuaries, wildlife, forests and wilderness, transportation, or land use—we are worse off than we were a decade ago. In spite of all the speeches, hearings, laws, and litigation, we have remained captives of old clichés. We are continuously brainwashed by press releases, by industrial advertising, and by public statements. Our priorities have been an overseas war, not the Three Hundred Year War at home. Population pressures mount; littering and pollution remain a scourge; the powerful lobbies seem bent on destroying our last few sanctuaries.

For things to change there must be a spiritual awakening. Our people—young and old—must become truly activist—

and aggressively so—if we and the biosphere on which we depend are to survive.

We can serve in that role only if we believe, with the Sioux, in "the goodness and the beauty and the strangeness of the greening earth, the only mother."[7]

SOURCE NOTES

THE THREE HUNDRED YEAR WAR

1 Parnall, Peter, *The Mountain,* New York: Doubleday, 1971.
2 Commoner, Barry, *The Closing Circle,* New York: Knopf, 1971, chapter 5.
3 *Ibid.,* chapter 9.
4 *Ibid.*
5 *Ibid.*
6 Linn, Alan, "Plankton Feeds the World," *Smithsonian,* March 1972.
7 Amos, William H., *The Infinite River,* New York: Random House, 1970.
8 Commoner, *op. cit.,* chapter 9.
9 Leopold, Aldo, *A Sand County Almanac,* New York: Oxford University Press, 1949, p. 204.

AIR

1 36 F.R. 8186
2 36 F.R. 15486
3 Nader, Ralph, *The Vanishing Air,* New York: Grossman Pub., Inc., 1970.

WATER

1 *Washington Monthly,* July 1971.
2 Reinemer, Vic, "Corporate Government in Action," *The Progressive,* November 1971.
3 Arthur E. Morgan in *Dams and Other Disasters,* Boston: Porter Sargent Publisher, 1971, gives many of the details.
4 Ridgeway, James, *The Politics of Ecology,* New York: E. P. Dutton, 1970.
5 Newfield, Jack, and Jeff Greenfield, *A Populist Manifesto,* New York: Praeger, 1972, p. 45.

RADIATION

1 *Congressional Record,* April 30, 1970, S6354.
2 U. S. Public Health Service, *Radiological Health Data and Reports,* April 1971, p. 189.
3 The arithmetic is in the July 8, 1971, *Congressional Record,* page S10653.
4 *Congressional Record,* April 30, 1970, S6342.
5 Gofman, John W., and Arthur R. Tamplin, *Population Control Through Nuclear Pollution,* Chicago: Nelson-Hall, 1971.
6 Curtis, Richard, and Elizabeth Hogan, *Perils of the Peaceful Atom,* New York: Ballantine Books, 1970.
7 Ford, Norman C., and Joseph W. Kane, "Solar Power," *Bulletin of Atomic Scientists,* October 1971.
8 The details are given in 117 *Congressional Record,* S16887 et seq.
9 84 Stat. 1566
10 Fabricant, Neil, and Robert M. Hallman, *Toward a Rational Power Policy,* New York: George Braziller, Inc., 1971.
11 Ridgeway, *op. cit.*
12 Kuznets, Simon, *Economic Growth of Nations,* Cambridge: Harvard University Press, 1971.
13 *The New York Times,* October 1, 1971.
14 *Environmental Action,* February 5, 1972, p. 11.

PESTICIDES

1 *Audubon,* November 1971, p. 5.
2 *The New York Times Magazine,* December 12, 1971, p. 38.

3 Shoecraft, Billee, *Sue the Bastards,* Phoenix: Franklin Press, 1971, p. 407.
4 Carson, Rachel, *The Silent Spring,* New York: Fawcett World Library, 1970.

GARBAGE

1 Fuller, Buckminster, *I Seem to Be a Verb,* New York: Bantam Books, 1970.

ESTUARIES

1 *Commissioner* v. *Volpe & Co.,* 349 MASS. pp. 104, 108.
2 *Id.,* p. 111.
3 430F (2d)199.
4 *Smithsonian,* March 1972, p. 40.

MINING

1 U. S. Department of Interior, *Surface Mining and Our Environment* 33 (1967) p. 63.
2 Berry, Wendell, *A Continuous Harmony,* New York: Harcourt, Brace Jovanovich, Inc., 1972.
3 Caudill, Henry, *Night Comes to the Cumberlands,* Boston: Atlantic Monthly Press, 1962.
4 "Strip-Mining Reclamation Techniques in Montana—A Critique," 32 Montana Law Review 65, p. 71.
5 Governor's Conference on Mined Land Reclamation and Montana Mining Law, Proceedings & Recommendations 37 (June 1970).
6 *Engineering & Mining Journal,* April 1967.
7 83 Stat. 852
8 Newfield, Jack, and Jeff Greenfield, *A Populist Manifesto,* New York: Praeger, 1972, pp. 45–46.

WILDLIFE

1 Olsen, Jack, *Slaughter the Animals, Poison the Earth,* New York: Simon & Schuster, 1971, pp. 140–41.
2 85 Stat. 480
3 *Field & Stream,* December 1971, p. 80.
4 Olsen, *op. cit.,* p. 280.

FOREST AND WILDERNESS

1 16 USC 528
2 U. S. Forest Service, 1971.
3 74 Stat. 215
4 *Ibid.*
5 Frome, Michael, *The Forest Service,* New York: Praeger, 1971.
6 78 Stat. 890
7 13 *Conserve,* November 1971, p. 4.
8 Stankey, George H., "Myths in Wilderness Decision Making," *Journn. Soil and Water Conservation,* September–October 1971.

TRANSPORTATION

1 70 Stat. 397
2 Nader, Ralph, *Politics of Law,* New York: Grossman Pub., Inc., 1971.
3 49 USC 1653 (f)
4 *Ibid.*
5 42 USC 4331
6 16 USC §460 (l) *et seq.*
7 16 USC §1241 *et seq.*

LAND USE

1 *Berman* v. *Parker,* 348 U.S. 26
2 Gov't. Code ¶51200–295
3 Gov't. Code ¶51201
4 Commoner, Barry, *The Closing Circle,* New York: Knopf, 1971, chapter 2.
5 *Audubon,* July 1971.
6 16 USC §460M *et seq.*
7 86 Stat. 44
8 Pub. Res. Code ¶¶4521–4618
9 3 ERC 1078, 1083
10 295 U.S. 495
11 42 USC 4331

POLITICAL ACTION

1 Reinemer, Vic, "Corporate Government in Action," *The Progressive,* November 1971.

2 *Field & Stream,* November 1971, p. 28.

3 83 Stat. 512, ¶4945. *General Explanation of Tax Reform Act of 1969* —prepared by the Staff of the Joint Committee on Internal Revenue Taxation, December 3, 1970, p. 49.

4 Sec. 16(c) . . . (4)

5 *Field & Stream,* December 1971, p. 74.

6 Skinner, B. F., *Beyond Freedom and Dignity,* New York: Knopf, 1971.

7 Armstrong, Virginia, *I Have Spoken,* Chicago: Swallow Press, 1971.

FEDERAL ENVIRONMENTAL AGENCIES

Air Pollution Control Service
Environmental Protection Agency
Room 3105, Waterside Mall
Washington, D. C. 20460

Atomic Energy Commission
Washington, D. C. 20545

Bureau of Indian Affairs
Department of Interior
Washington, D. C. 20240

Bureau of Land Management
Department of Interior
Washington, D. C. 20240

Bureau of Mines
Department of Interior
Washington, D. C. 20240

Bureau of Outdoor Recreation
Department of Interior
Washington, D. C. 20240

Bureau of Reclamation
Department of Interior
Washington, D. C. 20240

Bureau of Sport Fisheries & Wildlife
Department of Interior
Washington, D. C. 20240

Coast Guard
Department of Transportation
400 7th Street, S. W.
Washington, D. C. 20591

Corps of Army Engineers
Department of Defense
Washington, D. C. 20314

Council on Environmental Quality
722 Jackson Place, N. W.
Washington, D. C. 20006

Department of Labor
Occupational Safety & Health Administration
14th & Constitution Ave. N. W.
Washington, D. C.

Department of Transportation
Federal Highway Administration
400 7th Street, S. W.
Washington, D. C. 20590

Environmental Protection Agency
Room 3105
Waterside Mall
Washington, D. C. 20460

Federal Aviation Administration
Department of Transportation
800 Indiana Avenue, S. W.
Washington, D. C. 20591

Federal Power Commission
411 G Street, N. W.
Washington, D. C. 20426

Forest Service
Department of Agriculture
South Building
12th & Independence S. W.
Washington, D. C. 20250

Geological Survey
Department of Interior
Washington, D. C. 20240

Health, Education & Welfare
330 Independence Avenue, S. W.
Washington, D. C. 20201

Maritime Commission
Department of Commerce
14th & Constitution N. W.
Washington, D. C. 20230

National Bureau of Standards
Gaithersburg, Maryland
Mailing Address: Washington, D. C. 20760

National Marine Fisheries Service
Department of Commerce
Washington, D. C. 20230

National Oceanic and Atmospheric Administration
Department of Commerce
6001 Executive Blvd.
Rockville, Maryland 20852

National Park Service
Department of Interior
Washington, D. C. 20240

Office of Radiation
Environmental Protection Agency
3105 Waterside Mall
Washington, D. C. 20460

Office of Saline Water
Department of Interior
Washington, D. C. 20240

Rural Electrification Administration
Department of Agriculture
South Building
14th & Independence S. W.
Washington, D. C. 20250

Soil Conservation Service
Department of Agriculture
South Building
12th & Independence S. W.
Washington, D. C. 20250

Tennessee Valley Authority
Muscle Shoals, Alabama 35660

FEDERAL LEGISLATIVE COMMITTEES

SENATE

	Chairman
Judiciary Committee	James O. Eastland, Mississippi
Government Operations Committee	John L. McClellan, Arkansas
Select Committee on Small Business	Alan Bible, Nevada
Joint Committee on Atomic Energy	John O. Pastore, Rhode Island
Agriculture and Forestry Committee	Herman E. Talmadge, Georgia
Subcommittee on Environment, Soil Conservation and Forestry	James O. Eastland, Mississippi
Armed Services Committee	John C. Stennis, Mississippi
Subcommittee on Nuclear Test Ban Treaty Safeguards	Henry M. Jackson, Washington
Commerce Committee	Warren G. Magnuson, Washington
Subcommittee on Environment	Philip A. Hart, Michigan
Subcommittee on Oceans and Atmosphere	Ernest F. Hollings, South Carolina

Committee on Foreign Relations	J. W. Fulbright, Arkansas
Subcommittee on Oceans and International Environment	Claiborne Pell, Rhode Island
Committee on Interior and Insular Affairs	Henry M. Jackson, Washington
Subcommittee on Water and Power Resources	Clinton P. Anderson, New Mexico
Committee on Public Works	Jennings Randolph, West Virginia
Subcommittee on Air and Water Pollution	Edmund S. Muskie, Maine

HOUSE OF REPRESENTATIVES

	Chairman
Committee on Agriculture	W. R. Poage, Texas
Subcommittee on Conservation and Credit	W. R. Poage, Texas
Committee on Government Operations	Chet Holifield, California
Subcommittee on Conservation and Natural Resources	Henry S. Reuss, Wisconsin
Committee on Interior and Insular Affairs	Wayne N. Aspinall, Colorado
Subcommittee on Environment	Wayne N. Aspinall, Colorado
Committee on Interstate and Foreign Commerce	Harley O. Staggers, West Virginia
Subcommittee on Public Health and Environment	Paul G. Rogers, Florida
Committee on Merchant Marine and Fisheries	Edward A. Garmatz, Maryland
Subcommittee on Fisheries and Wildlife Conservation	John D. Dingell, Michigan
Subcommittee on Oceanography	Alton Lennon, North Carolina
Committee on Public Works	John A. Blatnik, Minnesota
Subcommittee on Flood Control and Internal Development	Robert E. Jones, Alabama

INDEX

About the Author

WILLIAM O. DOUGLAS was a practicing lawyer in New York City and the state of Washington, a law professor at Columbia and Yale universities and chairman of the Securities and Exchange Commission. He has been a member of the Supreme Court since 1939. Justice Douglas' hobbies include hiking, conservation, foreign travel and exploration. He is the author of thirty books, including: *Towards a Global Federalism, Russian Journey, Beyond the High Himalayas, Almanac of Liberty, Farewell to Texas, Points of Rebellion, International Dissent* and *Holocaust or Hemispheric Co-Op.* The present book reflects Justice Douglas' interest in legislation as well as his lifetime passion for the natural world.

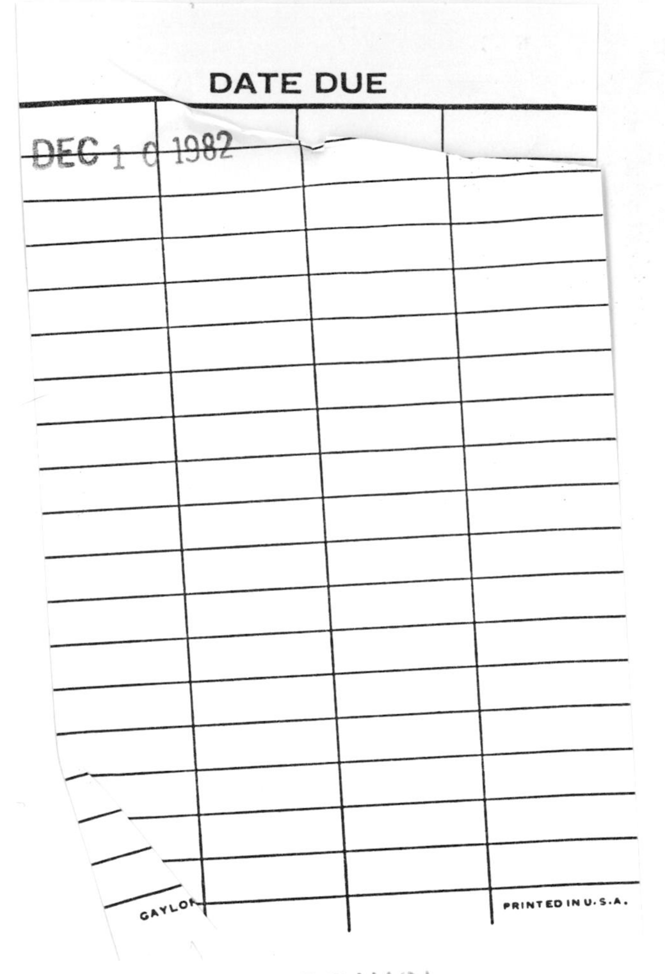

WITHDRAWN
UNIV OF MOUNT UNION LIBRARY